George Shaw

General Zoology
or
Systematic Natural History

Volume VIII

Part 2
Birds

Elibron Classics
www.elibron.com

Elibron Classics series.

ISBN 0-543-89996-9 (paperback)
ISBN 0-543-89995-0 (hardcover)

This Elibron Classics Replica Edition is an unabridged facsimile of the edition published in 1811 by Thomas Davison, London.

GENERAL ZOOLOGY,

or

SYSTEMATIC NATURAL HISTORY

by

GEORGE SHAW, M.D.F.R.S.&c.

WITH PLATES

from the first Authorities and most select specimens.

Engraved principally by

MRS GRIFFITH.

VOL. VIII. Part II.

AVES.

London, Printed for Kearsley, Wilkie and Robinson, Walker, Stockdale, Lea, Jeffry, Crosby, Longman Hurst Rees Orme & Brown, Scholey, Booth, Bagster, Gale & Curtis, Baldwin Lowe, and Davison.

1812.

GENERAL ZOOLOGY.

VOLUME VIII.——PART II.

BIRDS.

LONDON.
PRINTED FOR GEORGE KEARSLEY, FLEET STREET;
BY THOMAS DAVISON, WHITEFRIARS.

1811.

CONTENTS

OF

VOL. VIII.—PART II.

Directions for placing the Plates in vol. VIII. Part II.

The Vignette represents a diminished view of the yellow-collared Parrakeet, page 439.

* The Plate No. 79, SENEGAL CREEPER, should be 29, and placed opposite page 205, in Part I.

BIRDS.

ORDER

PICÆ.

RAMPHASTOS. TOUCAN.

Generic Character.

Rostrum maximum, inane, convexum, extrorsum serratum, apice subincurvatum.	*Bill* very large, light, hollow, convex, serrated outwards, slightly curved at the tip.
Nares pone rostri basin: orbitæ nudæ.	*Nostrils* behind the base of the bill: orbits naked.
Lingua angusta, elongata, pennacea.	*Tongue* long, narrow, feather-shaped.
Pedes scansorii.	*Feet* scansorial.

THE birds of this genus first became known to naturalists on the discovery of South-America, to the warmer regions of which they appear, in general, to be confined. Like the Hornbills, they are distinguished by the enormous size of their beaks, which, in some species, is nearly equal to that of the whole body. It is however of a very light substance, and, in the living bird, is even compressible between the fingers. Both mandibles are serrated in an outward direction: the tongue is of a highly singular form, representing the appearance of a very narrow lanceolate feather, being of a somewhat horny or cartilaginous nature, and divided on each side into innumerable short and close-set fibres; in consequence of which structure it was described by some of the old writers as a real feather, supplying the place of a tongue. The orbits of the eyes are generally bare. The Toucans are supposed to feed principally on fruits; but, in a state of captivity, have been observed to eat animal food of various kinds. They deposit their eggs, which are usually two in number, in the hollows of trees, on the surface of the decayed wood.

The species at present known are not very numerous, and of those usually considered as distinct, it is not improbable that some may be mere sexual differences.

M. Griffith sculp.

TOCO TOUCAN.

1811, Augt 1. London Published by G. Kearsley, Fleet Street.

TOCO TOUCAN.

Ramphastos Toco. *R. niger, collo antice uropygioque albis, crisso rubro, rostro rubente apice nigro.*

Black Toucan, with the fore-part of the neck and rump white, red vent, and reddish bill with black tip.

Ramphastos Toco. *R. nigricans, collo subtus uropygioque albis, orbitis lunula pectoris crissoque rubris. Lath. ind. orn.*

Le Toco. *Buff. ois.*

Toucan de Cayenne apellé Toco. *Pl. Enl.* 82.

Toco Toucan. *Lath. syn.*

THIS, which is one of the largest of the genus, measures about seventeen inches in total length, of which the bill alone measures about eight inches: it is of a pale rufous yellow colour, the base of both mandibles being black, and the tip of the upper mandible, to a considerable distance down the bill, of the same colour: the bird itself is also black, with a white throat, bounded beneath by a red bar across the breast: the rump is white, the vent-feathers red, and the legs black. It is a native of Cayenne, and seems to have been first described by Buffon.

YELLOW-BREASTED TOUCAN.

Ramphastos Tucanus. *R. niger, collo antice flavo, fascia pectorali crissoque rubris.*

Black Toucan, with the fore-part of the neck yellow, and red pectoral bar and vent-feathers.

Ramphastos Tucanus? *R. nigricans, fascia abdominali crisso uropygioque flavis. Lin. Syst. Nat.*

Tucana Surinamensis niger, ex albo, flavo, rubro mixtus. *Petiv. Gaz. t.* 44. *f.* 13.

Yellow-breasted Toucan. *Edw. pl.* 329. *Lath. syn.*

THIS, which measures about nineteen inches in length, is described by Edwards, from a living specimen brought at that time into England. "The bill, says Edwards, is very great in proportion, compressed sideways, having a sharp ridge along its upper part, and toothed on its edges: the upper mandible is green, with a long triangular spot of orange-colour on each side, and the ridge on the upper part yellow: the lower mandible is blue, with a shade of green in the middle: the point is red: it hath about five transverse faint dusky bars, which cross the joinings of the two mandibles: the nostrils are invisible, in the black line that surrounds the bill: the iris of the eye is of a fair green colour; round the eye is a broad space of naked skin, of a violet colour: the skin beneath the feathers is of a violet colour: the throat and breast are of a bright yellow; below which is a bar of scarlet feathers, which parts the yellow on the breast from the black on the belly: the covert-feathers of the tail are white above, and those beneath of a fine red: the crown of the head, upper part of the neck, the back, wings, belly, and tail are wholly black, though on the upper side of the wings and tail it has a changeable gloss of blueish purple: the legs and feet are all of a blue or violet colour."

The above specimen was brought from Jamaica, but the bird is a native of the hotter parts of South America.

BRASILIAN TOUCAN.

Ramphastos piscivorus. *R. nigricans, fascia abdominali crissoque rubris, uropygio albo. Lin. Syst. Nat. Lath. syn.*

Black Toucan, with red abdominal bar and vent-feathers, and white rump.

Xochitenacatl tertia. *Raii syn.*

Tucana Brasiliensis gutture albo. *Briss. av.*

Toucan, or Brasilian Pye. *Edw. pl. 64.*

Brasilian Toucan. *Lath. syn.*

This also is faithfully described, from a living specimen, by Edwards, who compares its size to that of a common tame pigeon, and its shape to that of a magpie, but with a larger head, and a much shorter tail in proportion.

"The bill, from the forehead, or the angles of the mouth, to the point, is six inches long; its height or depth, where deepest, is something more than two inches; from side to side, near the head, one inch thick: the upper mandible is of a pale yellow greenish colour: the sides, near the toothed edges, have each a long cloud of orange-colour, transversly barred with black or dusky lines, which pass through the divisions of the bill a little way into the sides of the lower mandible: the lower mandible is of an exceeding fine blue colour, faint towards the head, stronger towards the point: the point itself, of both upper and lower mandible, for above an inch depth, is of a fine scarlet colour: the bill is very much compressed sideways, and ends in a perfect edge along its upper part: the edge of the lower mandible is somewhat less ridged: the bill is bowed towards the point, and has a large cavity above the roof of the mouth: the nostrils are invisible, being situated pretty near together in the upper part of the bill; just in the line which passes between the bill and the forehead: the eyes are of a dark hazel-colour, encompassed round with a bare skin of a greenish yellow colour, having seams or creases running in a broken manner round the eye, and, reaching to the bill, there passes all round the basis of it a narrow black line: the top of the head, upper side of the neck, back, wings, belly, thighs, and tail, are of a black colour, the wings only having a shining changeable lustre: the sides of the head, throat, and breast, are white, or rather cream-colour; and between the white on the breast and the black belly, there is a crescent of fine red, whose horns point upwards on the sides, and this is softened both above and beneath into the colours it joins with: the rump, or coverts on the upper side of the tail, are white: the feathers beyond the vent, and those that cover the under side of the tail, are of a pale red: the legs, feet, and claws are of a light blue or violet colour."

Edwards adds, that, after death, the colours of the bill were totally lost or obscured, and the bare space round the eyes turned black.

This bird is a native of Brasil, and is, in all probability, no other than the female of the immediately preceding, from which it differs in having the breast cream-colour instead of yellow.

RED-BREASTED TOUCAN.

Ramphastos pectoralis. *R. niger, collo antice luteo, fascia lata pectorali crisso uropygioque rubris.*

Black Toucan, with the fore-part of the neck yellow, a broad, red pectoral band, and red vent and rump.

Ramphastos Tucanus? *R. nigricans, fascia abdominali, crisso, uropygioque flavis. Lin. Syst. Nat.*

Ramphastos dicolorus? *R. nigricans, pectore abdomine crisso uropygioque rubris. Lin. Syst. Nat.*

Toucan a gorge jaune. *Buff. 7. p.* 118.

Toucan a gorge jaune du Bresil. *Buff. 7. p.* 119.

Toucan a gorge jaune de Cayenne.
Toucan a gorge jaune du Bresil. } *Pl. Enl.* 269. 307.

Yellow-throated Toucan.
Yellow-breasted Toucan. } *Lath. syn.*

Length about eighteen inches: colour black, with a gloss or cast of green: cheeks, throat, and fore-part of the breast, in some sulphur-yellow, in others orange-yellow: across the lower part of the breast a broad crimson bar, sometimes extending nearly to the thighs, and sometimes falling far short of those parts; according to which variation the belly appears either black or crimson: thighs black: vent-feathers crimson; rump either crimson or orange-yellow: bill blackish olive-green, with pale yellow base bounded by a black bar: legs dusky. Native of Brasil and other parts of South America.

ALDROVANDINE TOUCAN.

Ramphastos Aldrovandi. *R. niger, pectore flavo, femoribus abdomine imo caudæque apice rubris.*

Black Toucan, with yellow breast, and red lower belly, thighs, and tail-tip.

Ramphastos picatus. *R. nigricans, pectore luteo, crisso rectricumque apicibus rubris, uropygio nigro. Lin. Syst. Nat.*

Pica Brasiliensis. *Aldrov. av. lib.* 12. *cap.* 19.

Brasilian Pie of Aldrovandus. *Will. orn. p.* 128. *pl.* 20.

The Brasilian Pie Toucan. *Albin. pl.* 25.

Le Toucan a ventre rouge. *Buff. ois.*

Preacher Toucan. *Lath. syn.*

Total length about twenty inches; of which the bill measures six, and is of a yellowish green colour with a reddish tip: the general colour of the bird is black, with a gloss of greyish green on the wings: the breast yellow or orange-colour; the thighs, lower part of the belly, vent-feathers, and tip of the tail bright red. This species is a native of Guiana and Brasil, and seems to have been first distinctly described by Aldrovandus. It is said to have a frequent habit of moving its head from side to side, while uttering its notes, which re-

RED BILLED TOUCAN.

1812, Aug. 1 London, Published by G. Kearsley, Fleet Street.

semble ill articulated words, and hence, in its native regions, has obtained the popular name of the Preacher.

RED-BILLED TOUCAN.

Ramphastos erythrorynchos. *R. niger, collo antice albo, fascia pectorali crissoque rubris, uropygio flavo.*

Black Toucan, with the fore-part of the neck white, red pectoral bar and vent-feathers, and yellow rump.

Ramphastos erythrorynchos. *R. nigricans, genis colloque subtus albis, fascia pectoris crissoque coccineis, uropygio sulphureo. Lath. ind. orn.*

Toucan à gorge blanche. *Buff. ois.*

Toucan à gorge blanche de Cayenne, appellé Tocan. *Pl. Enl.* 262.

Red-beaked Toucan. *Edw.* 238. *Lath. syn.*

First correctly described by Edwards. "This bird, says he, is of the bigness of a common-sized tame pigeon, and shaped like a Jackdaw: the bill is very large in proportion to the body; the head is also large in order to support it: the bill, from the angles of the mouth to its point, is in length six inches and a half; its height or width, in the thickest part, is a little more than two inches: its thickness near the head is one inch and a quarter: it is arched or rounded along the top of the upper mandible; the under side is also round: the upper mandible round its base, or joining to the head, and at its upper part quite to its point, is of a bright yellow-colour: its sides are of a fine red or scarlet colour, and so is the lower mandible, except at its base, which is purplish: the red, both on the upper and under chap, is clouded more or less in different parts with black; so that the point of the lower mandible is black: a black list passes almost round the bill near its base, which separates the red from the other colours: between the head and bill there passes a narrow black line, of separation all round the base of the bill, in the upper part of which the nostrils are placed, which do not shew themselves, being almost covered with feathers, which occasioned our first natural historians to say it was without nostrils, and set them on straining their wits to supply that want some other way: round the eyes, on each side of the head, is a space of blueish skin void of feathers, above which the head is black, except a white spot on each side, joining to the base of the upper mandible of the bill: the hinder part of the neck, the back, wings, tail, belly, and thighs are black: the under side of the head, the throat, and beginning of the breast, are white: between the white on the breast and the black on the belly is a space of red feathers in form of a new moon, having its points upwards: the covert-feathers under the tail are red, and those above the tail yellow: the legs, feet, and claws are of an ash-colour."

In the above specimen the tail, which is black, was wanting; but Edwards supplied that part to his own figure from a drawing of the same bird by Madam Merian, preserved in the British Museum. The Red-billed Toucan, like the rest of its tribe, is a native of South America, and particularly of Brasil and Guiana.

COLLARED TOUCAN.

Ramphastos torquatus. *R. supra niger, collo subtus albido, abdomine viridi postice rubro, torque rubro. Lath. ind. orn.*

Black Toucan, with the neck whitish beneath, green abdomen, red at the lower part, and red collar behind the neck.

Cochitenacatl. *Fern. Hist. Nov. Hisp. p. 46.*

Le Cochicat. *Buff. ois.*

Collared Toucan. *Lath. syn.*

DESCRIBED by Brisson and Buffon: total length eighteen inches; of the bill seven: upper mandible whitish; lower black: general colour of the bird black, with the back of the neck crossed by a red collar or bar: fore-part of the neck whitish, spotted with red, and streaked with black: belly green; vent-feathers red: thighs purple, and legs greenish. Native of those parts of Mexico which border on the sea, and supposed to feed on fish.

PAVONINE TOUCAN.

Ramphastos pavoninus. *R. corpore viridi rubro pavoninoque colore variegatus, rostro luteo nigroque vario. Lath. ind. orn.*

Toucan, with the body variegated with green, red, and peacock-colour, and bill varied with yellow and black.

Xochitenacatl. *Fernand. Hist. N. Hisp. p. 51.*

Hochicat. *Buff. ois.*

Tucana Mexicana viridis. *Briss. av.*

Pavonine Toucan. *Lath. syn.*

Described by Fernandez, and from him by Brisson and Buffon. Size that of the Brasilian Toucan: bill nearly four inches long: colour of the bird green, with a mixture of red and changeable or peacock-coloured feathers interspersed: legs black: native of the littoral parts of Mexico.

GREEN TOUCAN.

Ramphastos viridis. *R. nigro-virescens, capite colloque nigris (feminæ castaneis,) abdomine flavo, uropygio rubro, cauda cuneata.*

Blackish-green Toucan, with black head and neck, (in the female chesnut,) yellow abdomen, red rump, and cuneated tail.

Ramphastos viridis. *R. viridis, abdomine flavo, uropygio rubro. Lin. Syst. Nat.*

Toucan verd de Cayenne. *Buff. ois. Pl. Enl.* 727.728.

Tucana Cayanensis viridis. *Briss. av.*

Green Toucan. *Lath. Syn.*

Total length about fourteen inches: of the bill near three inches: upper mandible yellow at the

GREEN TOUCAN.

1811, Aug.t 1 London, Published by G. Kearsley, Fleet Street.

ARACARI TOUCAN.

1811, Augt 1. London, Published by G. Kearsley, Fleet Street.

top, and reddish on the sides, with a line of black separating the two colours: lower mandible black; with reddish or yellowish base: head and neck deep black: back, wings, and tail, dull or blackish-green, the larger quill-feathers and tail being nearly black: breast, belly, and vent-feathers pale yellow: rump red: tail slightly lengthened, and cuneated: legs and feet grey. In the female the head and neck are chesnut instead of black: native of Cayenne.

ARACARI TOUCAN.

Ramphastos Aracari. *R. nigro-virescens, abdomine flavo, fascia abdominali media uropygioque rubris, cauda cuneata.*

Blackish-green Toucan, with yellow abdomen, red middle abdominal bar and rump, and cuneated tail.

Ramphastos Aracari. *R. viridis, fascia abdominali, crisso uropygioque rubris, abdomine flavo. Lin. Syst. Nat.*

Aracari. *Marcgr. Bras. Will. orn. Ray. syn.*

Le Grigri. *Buff. ois.*

Toucan verd du Bresil. *Pl. Enl.* 166.

Aracari Toucan. *Lath. syn.*

Total length near seventeen inches: of the bill about four inches: upper mandible white, marked along the top by a longitudinal black stripe: lower mandible black: head and neck deep black, with a chesnut-coloured spot behind each ear: back, wings, and tail dull green: breast, belly, and sides pale yellow, the upper part of the breast reddish, and across the middle of the belly a moderately broad red bar: rump of the same colour: tail lengthened and considerably cuneated: legs and feet greenish. This species is known to vary, in having the wings and tail black, a black roundish spot on the breast, and a slight bar of the same colour on the abdomen. In the female the head and neck are brown instead of black. Native of South America, particularly of Brasil and Surinam.

PIPERINE TOUCAN.

Ramphastos piperivorus. *R. olivaceus, capite collo pectoreque nigris, macula postoculari lunulaque cervicali fulvis, cauda cuneata.*

Olivaceous Toucan, with black head, neck, and breast, orange ear-spot and nuchal crescent, and cuneated tail.

Ramphastos piperivorus. *R. viridis antice niger, crisso femoribusque rubris. Lin. Syst. Nat.*

Green Toucan. *Edw. pl.* 330.

Le Koulik. *Buff. ois.*

Toucan à collier de Cayenne. *Pl. Enl.* 577.

Toucan à ventre gris de Cayenne. *Pl. Enl.* 729.

Piperine Toucan. *Lath. Syn.*

THIS species has been described by Edwards, under the name of the Green Toucan. Its total length is about seventeen inches, and its general colour dull green: the head, neck, and breast deep black: behind each ear is a somewhat lengthened orange-coloured spot, and immediately behind the lower part of the neck an orange-coloured bar or collar: the belly is pale yellowish-green;

the thighs purple; the vent red, and the tail, which is cuneated, slightly tipped with dull red: the bill is about three inches long, and of a black colour, but towards the base varied with red, whitish, and orange-yellow. In the female the head, neck, and breast are brown instead of black, and the lower part of the belly grey instead of greenish-yellow. Native of Cayenne, and said to feed much on pepper; by which must be understood Capsicum or Cayenne pepper. The female of this species has been known to vary in having the bill horn-colour, with a black bar near the end, and two others near the edge; the ridge also being black, and the yellow crescent at the neck wanting.

BLACK-BILLED TOUCAN.

Ramphastos lutens. *R. flavescens, lateribus colli utrinque striga longitudinali nigra, cauda et alis albo nigroque variis, tectricibus alarum minoribus flavis. Lath. ind. orn.*

Yellowish Toucan, with a longitudinal black streak on each side the neck, wings and tail varied with black and white, and the smaller wing-coverts yellow.

Alia Xochitenacatl. *Will. orn. Nieremb. p.* 207.

Aracari à bec noir. *Buff. ois.*

Black-billed Toucan. *Lath. syn.*

Described by Buffon from Nieremberg, &c. Size that of a pigeon: bill black, thick, and bent: general colour of the body yellowish-white: on each side of the neck, from the bill to the breast, a black stripe: smaller wing-coverts yellow; remainder of the wings varied with black and white: tail of the same colour: legs brown. Native of Mexico.

BLUE TOUCAN.

Ramphastos cæruleus. *R. cæruleus cinereo variegatus, rostro corpore longiore, iridibus fulvis. Lath. ind. orn.*

Blue Toucan, varied with ash-colour, with the bill longer than the body, and fulvous irides.

Xochitenacatl. *Fern. Hist. N. H. p.* 47.

L'Aracari bleu. *Buff. ois.*

Blue Toucan. *Lath. syn.*

Described by Fernandez: size that of a pigeon: colour blue, intermixed with ash-colour: bill very large, and even longer than the body of the bird. Native of Mexico.

BLUE-THROATED TOUCAN.

Ramphastos dubius. *R. gutture cæruleo. Lath. ind. orn.*

Toucan with blue throat.

Blue-throated Toucan. *Lath. syn.*

This is mentioned by Dr. Latham from a catalogue of the Museum of Baron Faugeres of Montpellier, where it is announced as an undescribed species, but without any farther particulars relative to its colours or history.

SMOOTH-BILLED TOUCAN.

Ramphastos glaber. *R. corpore viridi, subtus virescente-flavo, capite colloque castaneis, uropygio rubro, rostro lateribus glabro. Lath. ind. orn.*

Toucan with green body, yellowish-green beneath, chesnut head and neck, red rump, and smooth-edged bill.

Smooth-billed Toucan. *Lath. syn.*

SIZE of the Green Toucan: length twelve inches: bill an inch and three quarters long, and one inch thick at the base; the upper mandible yellowish-brown, the lower black; the edges of both perfectly smooth or without serratures: orbits blue: head and neck chesnut, the top of the head darkest: upper parts of the body dark green; rump crimson: lower part of neck, breast, and belly pale greenish-yellow: thighs green, with the insides dusky: legs brown: described from Dr. Latham, and is, in all probability, the young or unadvanced state of the green or some other species. The two specimens from which Dr. Latham's description was drawn up were brought from Cayenne.

INDIAN TOUCAN.

Ramphastos Indicus. R. *niger viridi-cœruleo nitens, genis abdomineque albis, femoribus crissoque flavis, vertice uropygioque rubris.*

Black Toucan with blue-green gloss, white cheeks and abdomen, yellow thighs and vent, and red crown and rump.

Ramphastos Indicus. *Cimelia Physica. t. 57.*

This, which rests entirely on the faith of Mr. Miller's figure, is represented, perhaps through some error, as a native of India. Its total length is nearly twelve inches, and its general colour black, with blue-green glosses: the bill, which is perfectly smooth, and considerably less than in most others of the genus, is of an orange-red half-way from the base, the remainder being black: the base is also surrounded by a narrow black edge: the crown of the head is red; the cheeks white, and the orbits of the eyes naked and pale blue; the belly white, the thighs and vent yellow, and the upper tail-coverts red: the legs are black. This, like the immediately preceding, is probably an unadvanced specimen of some other species.

I cannot dismiss the present genus without observing that the true determination of the species is extremely difficult. The Count de Buffon considers the *Brasilian* and *Red-Billed* Toucans as no other than the female, (varying slightly in colour,) of the *Red-Breasted Toucan;* and consequently no less than four commonly supposed distinct species are by this author comprehended under one. Dr. Latham, though he describes them as distinct, seems yet inclined to believe that Buffon's supposition may be right. The *Red-Billed* Toucan however seems to differ so strikingly from the rest in the colour of the bill, that it can hardly be supposed the same with the Green-Billed or *Brasilian* Toucan; but it is not improbable that the *Brasilian* and *Yellow-Throated* Toucans of Edwards may constitute but a single species, of which the *Yellow-Throated* may be the male, and the *Brasilian* or White-throated the female.

SCYTHROPS. CHANNEL-BILL.

Generic Character.

Rostrum magnum, convexum, cultratum, longitudinaliter sulcatum, apice incurvo.
Nares rotundatæ, nudæ, ad basin rostri.
Lingua elongata, cartilaginea, apice bifida.
Pedes scansorii.

Bill large, convex, cultrated, longitudinally furrowed, and curved towards the tip.
Nostrils rounded, naked, at the base of the bill.
Tongue elongated, cartilaginous, bifid at the tip.
Feet scansorial.

AUSTRALASIAN CHANNEL-BILL.

Scythrops Australasiæ. *S. plumbeus, rectricibus albo nigroque fasciatis.*
Lead-coloured Channel-Bill, with the tail-feathers barred with black and white.
Scythrops Novæ Hollandiæ. *Lath. ind. orn.*
New Holland Channel-Bill. *Lath. syn. suppl. 2.*
Psittaceous Hornbill. *Phill. Bot. Bay. p. 165.*
Anomalous Hornbill. *White's Journ. p. 142.*

THIS genus, at present consisting of a single species only, is nearly allied to that of Ram-

AUSTRALASIAN CHANNEL BILL.

1811. Nov.r 2. London, Published by G. Kearsley, Fleet Street, & the other Proprietors

phastos, from which it principally differs in the greater strength or stoutness of the bill, and in having the tongue entire at the sides, and bifid at the tip.

The Australasian or New-Holland Channel-Bill is nearly of the size of a Crow, and measures in total length about seventeen inches, of which the bill measures almost four inches. The general proportions of the bird somewhat resemble those of a Cuckow, but with a longer and more cuneated tail. The colour of the upper parts of the body, wings, and tail is deep blueish ash-brown, the tips of the feathers somewhat more intense than the rest: the head, neck, and under parts of the bird are of a pale grey or dove-colour: the two middle tail-feathers have a black bar near the tip, which is white: all the remaining tail-feathers are ash-brown externally, but on the inner webs are white, crossed by numerous black bars, and marked, like the middle ones, by a broader black bar near the end, the tips being white: the eyes and the nostrils are seated in a reddish naked skin: the bill and legs are pale yellow; the former marked on the upper mandible by a longitudinal dusky streak or two, and on the lower by three or four dusky bars near the base.

This bird is a native of New-Holland, where it is sometimes seen in small flocks, but more frequently in pairs; frequenting trees, and uttering, during flight, a loud, screaming noise, not unlike the crowing of a cock. Its food appears to consist of fruit and insects.

CROTOPHAGA. KEEL-BILL.

Generic Character.

Rostrum compressum, arcuatum, dorso carinatum.	*Bill* compressed, arched, keeled on the top.
Nares perviæ.	*Nostrils* pervious.
Lingua compressa, apice subulata.	*Tongue* compressed, pointed at the tip.
Pedes scansorii.	*Feet* scansorial.

AMERICAN KEEL-BILL.

Crotophaga Americana. *Crotophaga nigro-violacea, viridi-nitens, cauda cuneata.*
Violet-black Keel-Bill, with green glosses and cuneated tail.
Crotophaga Ani. *C. pedibus scansoriis. Lin. Syst. Nat.*
Ani. *Marcgr. Bras. Will. orn.*
Razor-Billed Blackbird. *Catesb. Carol. append. pl. 3.*
L'Ani des Savanes. *Buff. ois.*
Petit Bout de Petun. } *Pl. Enl.* 102.
Grand Bout de Petun. }
Lesser and Greater Ani. *Lath. syn.*

THIS species is principally an inhabitant of the hotter regions of South America, and particularly of Brasil: it occurs however in various parts of North America, as well as in several of the West-

M. Griffith sculp.

AMERICAN KEEL BILL.

1811, Aug. 1. London, Published by G. Kearsley, Fleet Street.

Indian islands. There appear to be two races or varieties, agreeing in colour and general appearance, but differing in size; the larger being nearly of the size of a Magpie, and the smaller hardly exceeding that of a Blackbird.

In their mode of life these birds resemble the genus Buphaga, feeding principally on insects which infest cattle, and particularly on those of the genus Acarus ranking under the division Ricinus or Tick, which in those regions are often dreadfully injurious to the cattle: they also feed on various kinds of lizards, worms, caterpillars, &c. and, in defect of these, will often attack various kinds of vegetables, and particularly maize and rice. They generally frequent open places, and are never observed in woods of any considerable size. They usually fly in small flocks of twenty or thirty together, and when on the ground or perched, are always observed to carry their heads drawn in, or close to the shoulders, sitting near each other, and uttering a constant chattering cry, somewhat in the manner of Starlings. They are of a bold and fearless nature, and are scarcely alarmed at the sound of fire-arms, and as they are not numbered among edible birds, on account of the rankness of their flesh, they may be said to enjoy a kind of privileged security. Their wings are short and their flight feeble, and, during a storm, numbers are said to be destroyed. They breed in March, and build in thick bushes, forming a very large nest of interlaced twigs and grasses; lined with dried leaves. It is pretended that several females lay, and hatch their eggs in the same nest; so that these birds may be considered as in some degree analogous to Beavers among quadrupeds. The nest is about a foot and half in diameter, and of a very considerable depth; but varies according to the number of families that are to inhabit it. Some individuals however are said to build single nests, or else make divisions in the cavity of a large one for their own offspring: where this is not done, the whole of the eggs roll together in the middle of the nest, and one bird is said often to cover or sit on the whole; but it is to be supposed that, in this case, they relieve each other by turns: when the nest is at any time left without an attendant, care is taken to cover the eggs with a layer of leaves, both for concealment, and in order to preserve a proper degree of heat, and when the numerous family is hatched, the busy mothers are assisted by the males in feeding the young; and where several groupes have been hatched in the separate inclosures of the larger nests, the parents are said to feed all in common, without making any particular distinction between their own offspring and that of their neighbours. They are observed to breed several times in a year, and the eggs, which are several in number, are of a blueish-green colour.

The Keel-Bill is a bird of a tame and gentle nature, and when taken early, may be educated in the same manne as a Parrot, and taught to speak; but its disagreeable smell is a reason for seldom practising this experiment.

Whether the larger and smaller races above-mentioned ought to be considered as constituting one and the same species only, or whether they are truly distinct, can perhaps only be ascertained by an accurate examination of all particulars relative to these birds when observed in their native regions.

PSITTACUS. PARROT.

Generic Character.

Rostrum aduncum, mandibula superiore mobili, plerisque cera instructa.	*Bill* hooked, with the upper mandible moveable, and generally furnished with a cere.
Nares in rostri basi.	*Nostrils* in the base of the bill.
Lingua plerisque carnosa, obtusa, integra.	*Tongue*, in most species, fleshy, obtuse, entire.
Pedes scansorii.	*Feet* scansorial.

THIS most numerous and splendid genus is chiefly confined to the warmer regions of the ancient and new continent, or within the limits of the tropics, none being natives of Europe. Some few however are found in latitudes far beyond what was supposed by Buffon, and even as far as forty or forty-five degrees on each side the equator. They may be considered, in the Linnæan phrase, as the Monkeys of Birds, being remarkable for their active and imitative disposition. From the peculiar form of their tongue, which in most species is thick, flattish, rounded, and fleshy, they are often enabled to articulate with greater distinctness than other birds: the upper mandible is

M. Griffith sculp.

SKELETON OF ASH COLOURED PARROT.

1811, Aug. 1. London, Published by G. Kearsley, Fleet Street.

moveable, and the feet formed for climbing, with a power also of bringing forward at pleasure one of the hind toes. They are frugivorous and monogamous, depositing their eggs, which are generally two in number, in the holes of decayed trees. Though generally observed in pairs, they sometimes assemble in vast flocks.

The whole genus may be very properly divided, according to the example of Linnæus, into the *Psittaci macrouri*, or those which have more or less long, cuneated, and pointed tails, and into *Psittaci brachyuri*, or those which have more or less short and even-feathered tails. These divisions however are not to be understood as settled with rigorous exactitude, it being hardly possible to mark the precise limits of the two assortments.

With more or less long and pointed tails.

SCARLET MACCAW.

Psittacus Macao. *P. coccineus, alis cœruleis, tectricibus luteo-variis, genis nudis rugosis albis.*

Scarlet Maccaw, with blue wings, wing-coverts varied with yellow, and white naked wrinkled cheeks.

Psittacus Macao. *P. macrourus ruber, remigibus supra cœruleis, subtus rufis, genis nudis rugosis. Lin. Syst. Nat.*

L'Ara rouge. *Buff. ois. Pl. Enl.* 12.

Le petit Ara rouge? *Buff. ois. Pl. Enl.* 641.

Red and blue Maccaw. *Edw. pl.* 158. *Lath. syn.*

L'Ara Canga. *Levaill. Perr. pl.* 2.

L'Ara Macao. *Levaill. Perr. pl.* 1.

L'Ara tricolor? *Levaill. Perr. pl.* 3.

Psittacus Macao. *Lath. ind. orn.*

Psittacus Aracanga. *Lath. ind. orn.*

The Scarlet Maccaw, by far the most magnificent of the whole Parrot tribe, is a native of South America and of some of the larger West-Indian islands, where it resides in woods, and feeds, like all the rest of this vast and splendid genus of birds, on various kinds of fruit. Its plumage is accurately described by the ingenious Edwards, who selected for this purpose a specimen in the highest state of perfection. The best figure extant is also that of Edwards, which, in the true expression of character, far surpasses that given in the Planches Enluminées of Daubenton.

" This bird, says Edwards, is undoubtedly the first of the Parrot kind, if we consider either its

GREAT SCARLET MACCAW.

1808 Oct.1 London Publish'd by G.Kearsley Fleet Street.

magnitude, or the great variety and beauty of the colours with which its plumage is adorned. It is the biggest of all the kinds I have met with: when the tail is perfect, I have found some of them to measure more than thirty-six inches from bill-point to tail-end: the arch of the upper mandible of the bill, from the forehead to the point of the bill is near three inches: the leg, from the knee downwards, is not an inch and half long; the longest toe, with the claw, is two inches and a half long: the upper mandible of the bill is whitish, except on each side next the head, where it is dusky: the lower mandible is black or dusky: it hath not a bare skin, covering the bill, as some Parrots have: the nostrils are placed in the upper part of the bill, just within the feathers: the bill is great and strong; the tongue roundish and soft; the sides of the head, from the bill backwards, for a good broad space, are bare of feathers, and covered with a whitish, wrinkled, rough skin: in the upper parts of these spaces the eyes are placed, whose irides are yellow: the head, neck, breast, belly, thighs, upper part of the back, and lesser covert-feathers of the wings are of a very fine bright red or scarlet colour: the quill-feathers of the wings are of a very fine blue on their outsides, and a faint red on their under sides: the first feathers next above the quills are of a fine yellow colour, some of the feathers being tipped with green: the blue quills which fall next the back are tinged with green: the hinder part of the thigh has some green intermixed with the red:

the lower belly, and coverts under the tail, as also the lower part of the back, and coverts on the upper side of the tail, are of a very fine blue colour: the tail-feathers gradually shorten towards the sides; some of the longest or middle-feathers are wholly red; the shorter or side-feathers are partly red and partly blue; their tips being blue, and their bottoms red: the legs and feet are covered with blackish or dusky scales: the toes are disposed two forwards and two backwards, as in other Parrots, all armed with strong claws."

This noble bird, at its first introduction into Europe, was justly considered as a present fit for royal personages, and was one of the principal ornaments in the halls of palaces. It seems to have been extremely rare till towards the decline of the sixteenth century.

Aldrovandus, a zealous naturalist, mentions his having seen one at the court of the Duke of Mantua, where he also observed a very fine specimen of the next species, or Great Blue and Yellow Maccaw.

The Scarlet Maccaw is a bird which occasionally varies in some degree in point of size and colours, but the differences seem to be merely such as may be supposed to result from the more or less advanced age of the bird, and a more or less perfect state of plumage.

The manners of this species and the next or Blue and Yellow Maccaw, in a state of nature, are said to be exactly similar, and are well detailed in the Ornithology of the Count de Buffon. They live, according to the testimony of those who have studied their habits, in the woods that cover swampy grounds, and which abound in palm-trees, feeding on the fruit of the Palmettoes or Borassi, of which there are innumerable forests in the over-flowed savannahs of South-America. They generally appear in pairs, and but seldom in flocks: sometimes however they assemble together, and their united screams are then heard at a vast distance. Of all the Parrot tribe they fly the best, and are always observed to perch on the summits of trees, or on the highest branch. During the day they wander to the distance of about a league from their favorite spot or home, but always return in the evening. They build in the holes of old trees, which are very common in their native regions, and even more numerous than the rising and vigorous plants. They enlarge the hole when too narrow, and line the inside with feathers. They have two hatches annually, like all other American Parrots, and each consists of two eggs, which are said to be of the size of Pigeon's eggs, and spotted like those of a Partridge. The males and females sit alternately on the eggs, or cherish the young, and both equally carry the food: they never desert their charge, so long as their assistance is required, and always perch near their nest. The young are said to be easily tamed, and in many parts of South-America these birds are never taken but in the nest, the grown birds being much less easily educated. Fernandez relates that they may be taught to speak, but that their articulation is hoarse and unpleasant. It is but rarely however that those which are brought to Europe are known to articulate more than a word or two; and their general voice is a loud and piercing scream. In a state of captivity, like many other birds, they are subject to epileptic fits, which however do not prevent them from arriving at a very considerable age, sometimes not less than thirty years.

To this general description of the manners of the Maccaw it would be unpardonable not to add the short but striking description of a flight of Maccaws, which made its appearance in view of Lord Anson and some of his company, while seated in a romantic spot in the island of *Quibo*, and admiring the view of a magnificent waterfall.

"Near the north-east point of the island, says the relator, they discovered a natural cascade, which surpassed, as they conceived, every thing of this kind which human art or industry hath hitherto produced. It was a river of transparent water, about forty yards wide, which ran down a declivity of near a hundred and fifty yards in length: the channel it ran in was very irregular, for it was entirely formed of rock; both its sides and bottom being made up of large detached blocks, and by these the course of the water was frequently interrupted, for in some places it ran sloping, with a rapid but uniform motion, while in other parts it tumbled over the ledges of rocks with a perpendicular descent. All the neighbourhood of this

BLUE & YELLOW MACCAW.

1811, Aug. 1. London, Published by G. Kearsley, Fleet Street.

stream was a fine wood, and even the huge masses of rock which hung over the water, and which by their various projections formed the inequalities of the channel, were covered with lofty forest-trees. While the Commodore and those who were with him attentively viewing the place, were remarking the different blendings of the waters, the rocks, and the woods, there came in sight, as it were, still to heighten and animate the prospect, a prodigious flight of Maccaws, which hovering over this spot, and often wheeling and playing on the wing about it, afforded a most brilliant appearance, by the glittering of the sun on their variegated plumage; so that some of the spectators cannot refrain from a kind of transport when they recount the complicated beauties which occurred in this extraordinary water-fall."

BLUE AND YELLOW MACCAW.

Psittacus Ararauna. *P. supra cœruleus, subtus luteus, genis nudis, lineis plumosis nigris.*

Blue Maccaw, deep yellow beneath, with naked cheeks marked by black plumy lines.

Psittacus Ararauna. *P. supra cœruleus, subtus luteus, genis nudis, lineis plumosis. Lin. Syst. Nat.*

Ararauna. *Marcgr. Bras.*

Psittacus maximus cyano-croceus. *Aldr. Will.*

Blue and Yellow Maccaw. *Edw. pl.* 159. *Lath. syn.*

L'Ara bleu. *Buff. ois. Pl. Enl.* 36. *Levaill. Perr. pl.* 3.

THIS beautiful bird is but very little inferior in point of size to the preceding, which it also resembles in its general shape and proportions. It is described in the following manner by Edwards.

"The bill is arched, the upper part hooked and hanging over the nether, and is all of a black colour: the nostrils are placed at the base of the upper mandible, in a white bare skin, which extends itself on the sides of the head all round the eyes, and a good way beneath them: these white bare plats of skin on the sides of the head are variegated with fine lines of small black feathers, which appear like needle-work: the circle round the eye is of a pale yellow colour: it hath immediately under the bill a large black spot, which turns round and upwards on its sides, and encompasses part of the bare white space on the sides of the head: the feathers on the top of the head are green, which gradually become blue on the neck: the upper side of the neck, the back, and upper sides of the wings and tail are of an exceeding fine blue colour, with some little variation of shade, viz. the lesser coverts of the wings and the rump are a little tinctured with green, and the tail and quill-feathers on their upper sides a little tinctured with purple: all the blue feathers of the back, wings, and tail are of a reddish yellow on their under sides: the fore-part of the neck, the breast, belly, thighs, and covert-feathers under the tail, are of a fine yellow-orange-colour, except the hinder part of the thighs, where there is a little blue intermixed: the covert-feathers withinside the wings are yellow, which appears outwardly on the ridge or joint in the upper part of the wing: the legs and feet are of a blackish colour."

This species is less common than the Scarlet Maccaw, of which it has by some been most erroneously considered as the female. It seems to have been first described by Aldrovandus, from a specimen which he saw in the year 1572, in the palace of the Duke of Mantua. It was brought over from America in 1569, and lived upwards of twenty years. Its death was occasioned by the gout in its feet, which at length swelled, and at intervals were extremely painful; the bird expressing its sufferings at such times by most lamentable screams; and in one of these fits it died, in the month of February 1599.

HYACINTHINE MACCAW.

Psittacus augustus. *P. cyaneus, rostro pedibusque nigris, orbitis basique mandibulæ inferioris luteis.* *Museum Leverianum.* No. 2.
Deep-blue Maccaw, with the bill and legs black; the orbits and base of the lower mandible yellow.
Hyacinthine Maccaw. *Nat. Misc. vol.* 15. *pl.* 609.
Psittacus hyacinthinus. *P. macrourus violaceo-cæruleus, capite colloque dilutioribus, orbitis gulaque nudis flavis. Lath. ind. orn.*
Hyacinthine Maccaw. *Lath. syn.*

This august species, which is a degree larger than the Blue and Yellow Maccaw, was first described by myself, in the work entitled Museum Leverianum, from a very fine specimen preserved in that celebrated collection. I must therefore repeat my former description. The colour of this bird is the richest and deepest mazarine blue, uniformly diffused over the whole plumage; except that on the edges of the wings and the forehead is a very slight tendency to a sea-green or blueish-green cast: the under surface of the wings and tail is black: the beak is most uncommonly large and strong, and considerably exceeds those of the Macao and Ararauna: it is totally black: the legs and feet are also black, and extremely strong: the orbits or bare spaces round the eyes are of a deep yellow, and the base of the lower mandible is surrounded by a bare skin of the same colour: the general proportions of this magnificent bird are the same as in the two species above-mentioned. Nothing certain is known relative to its native country; but it may be presumed that, like the rest of the large Maccaws, it is an inhabitant of South-America. It may be added, that the specimen in the Leverian Museum was probably the first of its kind ever brought into Europe, and before its introduction into the Leverian Museum, had been in the possession of the then Lord Orford, with whom it lived a considerable time. I have been informed that a second specimen was some time afterwards brought to Lisbon, and was presented to the Queen of Portugal. It is therefore probable that the bird is of Brasilian origin. It is remarkable that it differs from the rest of the Maccaws in having the cheeks covered with feathers.

MILITARY MACCAW.

Psittacus militaris. *P. viridis fronte rubro, genis nudis, remigibus uropygioque cæruleis, rectricibus rubris apice cæruleis.*

Green Maccaw, with red forehead, naked cheeks, blue quill-feathers and rump, and red tail-feathers tipped with blue.

Psittacus militaris. *P. macrourus viridis, alis cæruleis, fronte caudaque rubris, genis nudis lineis plumosis. Lin. Syst. Nat. Lath. syn.*

Great Green Maccaw. *Edw. pl.* 313.

Le Grand Ara militaire. *Levaill. Perr. pl.* 6.

Military Maccaw. *Lath. syn.*

The first description of this species seems to be that of Edwards, who represents it as a Maccaw of the first magnitude, the wing, when closed, measuring about thirteen inches in length, and the middle tail-feathers about fifteen; the general proportions of the bird seeming to be similar to the species before described. "The head, says Edwards, was large in proportion to the bird: the bill was of a dusky colour, pretty strong, the upper mandible pretty much hooked, having angles on each side: the tongue dusky, round, and soft: the nostrils small, placed in a narrow whitish skin that wholly surrounded the bill: on each side of the head was a pretty broad portion of skin, bare of feathers, of a flesh-colour, with several oblique lines under the eyes, composed of minute black feathers: the eyes are placed in these bare parts of the head: their irides are bright yellow; the pupils black: the forehead, adjoining to the bill, is covered with beautiful red feathers: a little dusky red is seen under the bill, mixing with the green feathers: the top of the head, whole neck, breast, and feathers of the wings, are of a fine full green colour: the green coverts beneath the tail are a little mixed with red: the quills and part of the row of coverts above them are of a very fine sky-blue, except a few next the back, which gradually become green: the insides of the wings, and the under side of the tail are of a dirty orange-colour: the middle of the back, the rump, and coverts of the tail are of a fine blue: the middle feathers of the tail are very long, and gradually shorten towards the side-feathers, which are not above a third the length of the middlemost; they are all pointed, and of a fine red colour, with blue tips: their number is twelve: the legs and feet are covered with scales of a dusky flesh-colour; the claws dusky."

The native region of the above bird was unknown to Edwards, but it is since ascertained to be an inhabitant of Guiana. A variety or smaller race also occurs, and which appears to differ in size only from the larger; and this, according to Monsr. Levaillant, has been confounded by the Count de Buffon with the following species.

Monsr. Levaillant figures also a variegated specimen, patched with red, yellow, &c. an accident common, as he observes, to almost all the Parrot tribe; and with respect to the supposition, somewhere mentioned by Buffon, of producing colours ad libitum by means of art, he very justly considers the notion as totally absurd, unless by absolutely dying or painting the feathers.

BRASILIAN GREEN MACCAW.

Psittacus severus. *P. viridis, genis nudis, fronte fusco-purpureo, remigibus rectricibusque cæruleis subtus fusco-rubentibus.*

Green Maccaw, with naked cheeks, purple-brown front, and blue wing and tail-feathers dusky-red beneath.

Psittacus severus. *P. macrourus viridis, genis nudis, remigibus rectricibusque cæruleis subtus purpurascentibus. Lin. Syst. Nat.*

Brasilian Green Maccaw. *Edw. pl. 229. Lath. syn.*

Maracana. *Marcgr. Bras. Will. orn.*

L'Ara vert. *Buff. ois.*

L'Ara Marakana. *Levaill. Perr. pl. 8. 9.*

L'Ara verd du Bresil. *Pl. Enl. 383.*

General length about seventeen inches: size that of a tame pigeon: habit similar to that of the Military Maccaw: colour fine green; the bend of the shoulders, and whole underside both of wings and tail red: quill-feathers and some of the larger coverts fine blue: tail green above, but growing blue at the tips; the two middle feathers blue throughout their whole length on the outer edges: bill black, with flesh-coloured cere: orbits pale flesh-coloured, with a few minute scattered black plumes: feathers round the bill, both above and beneath, blackish: irides yellow: legs black, with a feathery red zone round the bottom of the thighs: native of Brasil, where, according to Monsr. Levaillant, it is extremely common; appearing in innumerable flocks, and committing great devastation among the coffee-plantations, by devouring the ripe berries. The same author adds, that the old birds make an excellent soup, while the young are equally delicate when roasted. This species seems to have been first described by Marcgrave, under its native name *Maracana*. Monsr. Levaillant observes that the figure given in the Planches Enluminées, No. 383, instead of a narrow dusky or blackish-purple bar on the forehead, has a broad red one, like that of the former species; and this, he observes, has been evidently done in order to accommodate the plate to Buffon's description, who confounded it with the preceding species.

PARROT MACCAW.

Psittacus Makavouanna. *P. viridis, genis nudis albis, subtus subrufescens, vertice remigibusque cærulescentibus.*

Green Maccaw, with naked white cheeks, beneath subrufescent, with blueish crown and wing-feathers.

Psittacus Makavouanna. *P. macrourus viridis, genis nudis, subtus rufescens, abdomine virescente, remigibus cæruleis, apicibus extus fuscis. Lath. ind. orn.*

Perriche Ara. *Buff. ois.*

La Perruche Ara de Cayenne. *Pl. Enl.* 864.

L'Ara Macavouanne. *Levaill. Perr. pl.* 7.

Parrot Maccaw. *Lath. syn.*

Described by Barrere in his *France Equinoxiale*, and from him by Buffon. Length eighteen inches; of which the tail measures nine inches: colour deep brownish green above, except the larger quill-feathers, which are blue, edged with green, and tipped with brown: head green, with a cast of deep blue, appearing blue in some lights: cheeks bare and whitish: throat and breast tinged with rufous: remainder of the under parts of a paler green than the upper: lower part of the abdomen, and vent-feathers, reddish brown: under parts both of wings and tail yellowish-green: bill and legs blackish. Native of Cayenne, where it is considered as a migratory bird. It frequents the overflowed savannahs, and feeds on the fruits of the Palmetto. Its native name is *Makavouanne*.

BLACK MACCAW.

Psittacus ater. *Psittacus macrourus nigricans, viridi-splendidissimus, rostro oculisque rubentibus, pedibus flavis. Lath. ind. orn.*
Blackish Maccaw, with a strong gloss of green, reddish eyes and bill, and yellow legs.
Ararauna, ou Machao. *De Laet. descr. des Ind. occ. p. 490.*
Ara Noir. *Buff. ois.*
Black Maccaw. *Lath. syn.*

Described by Buffon, but merely from report: said to be a native of Guiana, and supposed by Buffon to be the same with the *Ararauna* or *Machao* of De Laet in his *Description des Indes Occidentales,* and which is said to be of a black colour, but so blended with green as to appear very splendid in the sunshine: bill and eyes reddish: legs yellowish. It is said to differ from the rest of the Maccaws in never approaching the settlements of the natives; remaining altogether among the rocks and mountains.

OBSCURE MACCAW.

Psittacus obscurus. *P. macrourus fuscus, genis nudis rubris, vertice cinereo-nigresente vario, cauda cinerea. Lin. Syst. Nat. Lath. ind. orn.*

Dusky Maccaw, with naked red cheeks, crown varied with ash-colour and black, and cinereous tail.

Psittacus obscurus. *Hasselq. itin. p. 236.*

This species seems obscure in every sense of the word; being known only from the short description given by Hasselquist, and repeated by Linnæus. It is said to be of the size of a Jay or a Cuckow, with a black bill, surrounded by black rough feathers, intermixed with hairs: the crown of the head variegated with black and grey: the upper parts of the neck and wings black: the belly and thighs ash-coloured, with transverse hoary lines: the legs furnished with black tubercles; the claws black; the tail long and cuneiform. Native of Africa.

NOBLE MACCAW.

Psittacus nobilis. *P. macrourus viridis, genis nudis, humeris coccineis. Lin. Syst. Nat.*
Green Maccaw, with naked cheeks, and scarlet shoulders.
Psittacus nobilis. *Lin. Mus. Ad. Frid. 2. p. 13.*
Noble Parrot. *Lath. syn.*

KNOWN only from the short description given by Linnæus in his work entitled *Museum Adolphi Friderici*, where it is said to be of the size of a Turtle, and of a green colour, with the bend of the wing scarlet; the face naked and white; and the tail cuneated. Native of Surinam.

PAVOUANE PARRAKEET.

Psittacus Guianensis. *P. viridis, orbitis nudis albidis, costa humerali tectricibusque inferioribus rubris, remigibus rectricibusque subtus flavescentibus.*

Green Parrakeet, with naked whitish orbits, ridge of the shoulders and under wing-coverts red, and quill and tail-feathers yellowish beneath.

Psittacus Guianensis. *P. macrourus viridis, genis rubro maculatis, tectricibus alarum minoribus inferioribus coccineis, majoribus luteis, armillis rubris. Lath. ind. orn.*

La Perriche Pavouane. *Buff. ois.*

Perruche de la Guiane. *Pl. Enl.* 167. 407.

La Perruche Ara Pavouane. *Levail. pl.* 14.

Pavouane Parrot. *Lath. syn.*

Described by Brisson and Buffon as of the length of twelve inches: head and upper parts fine green; cheeks and sides of the neck, in the old birds, speckled with bright red; gradually becoming more conspicuous as the bird advances in age: smaller wing-coverts red; greater yellow, and both quill and tail-feathers dusky yellow beneath: bill whitish; legs and feet grey. Native of Cayenne and the Antilles, where it is not uncommon; often flying in flocks; frequenting wooded savannahs, and feeding in preference on the berries of the Erythrina Corallodendron. Buffon observes, that in one of the plates of the Planches Enluminées, viz. pl. 167, the bill is erroneously represented of a red colour.

Monsr. Levaillant observes that the Pavouane

Parrakeet varies considerably both in size and colour according to the regions in which it occurs. In Guiana it is not only smaller, but less brilliant than in the Antilles, where its length, according to Monsr. Levaillant's figure, seems to be near fourteen inches. The red spots on the cheeks, described by Buffon as appearing in the old birds, and the red garter round the thighs, mentioned by Brisson, are, as Monsr. Levaillant assures us, merely owing to that variegation of plumage which occasionally takes place in all the Parrot tribe, when in a state of captivity, and are by no means to be considered as forming any part of the natural character of the species.

It may be here not improper to observe, that Monsr. Levaillant makes a very important observation relative to the long-tailed Parrakeet tribe in general; viz. that though this tribe may properly enough be allowed to constitute two natural subdivisions, one distinguished by having the tail regularly or gradually cuneated by the successive elongation of the side-feathers to the middle ones, and the other by an elongation of the side feathers to a certain distance only, the two middle ones running out to a great extent beyond them, yet in a state of domesticity it not unfrequently happens, in consequence of the process of moulting, that the genuine shape of the tail is injured or altered in its proportions; thus causing a great degree of uncertainty as to the tribe to which the bird belongs. It is to this circumstance that Monsr. Levaillant attributes the mistakes in the work of Buffon, who has often described the same species under different names, and under different divisions in the tribe. Hence the necessity of obtaining, if possible, such specimens as have been taken in their truly natural or wild state.

BANDED PARRAKEET.

Psittacus vittatus. *P. viridis, subtus olivaceo-flavescens, orbitis nudis, fronte fusco-purpureo, femoribus caudaque subtus rufescentibus.*

Green Parrakeet, yellowish-olive beneath, with naked orbits, purple-brown front, reddish thighs, and tail reddish beneath.

La Perruche Ara a bandeau rouge. *Levaill. pl.* 17.

General appearance much allied to that of the immediately preceding, but of rather smaller size, and of a more slender and elegant aspect: colour green, with olive throat, the feathers of which are edged with pale yellow: remainder of the under parts pale olive, edged in a similar manner: thighs, and under surface of the tail rufous: across the forehead, immediately beyond the base of the upper mandible, a narrowish purple-brown bar, slightly speckled with red: bill and legs pale. Native of Brasil, and considered by Levaillant as a species before undescribed.

WAVE-BREASTED PARRAKEET.

Psittacus versicolor. *P. viridis, collo subtus pectoreque nigricantibus aurantio-undulatis, abdomine medio caudaque subtus rubris.*

Green Parrakeet, with the fore-part of the neck and breast blackish, waved with orange-colour, the middle of the belly and under surface of the tail red.

Psittacus versicolor. *P. macrourus viridis, capite corporeque subtus fuscis, pennis juguli aureo, abdominis cæruleo-rubro undatis. Lath. ind. orn.*

La Perriche a gorge variée. *Buff. ois.*

Perruche a gorge tachetée de Cayenne. *Pl. Enl.* 144.

La Perruche Ara a gorge variée. *Levaill. pl.* 16.

Waved-Breasted Parrakeet. *Lath. Syn.*

VAR.

Psittacus squammosus. *P. macrourus viridis, capite collo pectoreque marginibus pennarum aurantiis, humeris coccineis, uropygio abdomineque medio sanguineis. Lath. ind. orn.*

Scaly-Breasted Parrakeet. *Lath. syn.*

An elegant species: size rather smaller than that of a Blackbird: length about ten inches: colour fine deep green: forehead blue; cheeks orange-brown: back part of the head dusky: throat, fore-part of the neck, and breast blackish, each feather edged with orange-colour: ridge of the shoulders, lower part of back, and rump red: outside edges of the quill-feathers blue, with dusky tips: middle of the belly red, or ferruginous with dusky undulations: thighs green: under surface of the tail red or ferruginous: bill and legs dusky: orbits naked and whitish.

The variety, described as a distinct species by Dr. Latham, under the name of *Scaly-Breasted Parrakeet*, seems to differ chiefly in having the head itself, as well as the fore-part of the neck and breast coloured as in the first-described kind, and in having the lower part of the back and rump, and the middle of the belly blood-red.

This species is a native of Cayenne, and is considered by Buffon as a very rare Parrakeet. It is however, according to Monsr. Levaillant, very common throughout all Guiana.

EMERALD PARRAKEET.

Psittacus smaragdinus. *P. viridis, abdomine subviolaceo, crisso caudaque rubro-purpureis.*

Green Parrakeet, with subviolaceous abdomen, and purple-red vent-feathers and tail.

Psittacus smaragdinus. *P. macrourus viridis nitens, abdomine postico caudaque ferrugineo-castaneis. Lath. ind. orn.*

Perruche emeraude. *Buff. ois. Levaill. pl.* 21.

Perruche des terres Magellaniques. *Pl. Enl.* 85.

Emerald Parrot. *Lath. syn.*

Length rather more than thirteen inches: colour fine glossy green, the belly tinged with violet, and the vent-feathers and tail dusky-red or purple, darker beneath: the feathers on the whole bird have narrow dusky edges, as in many of the Parrot tribe, giving a scaled appearance to the plumage: bill and legs black. Native, according to report, of the Magellanic regions; but Buffon is not will-

TABUAN PARRAKEET.

1811 Aug.t 1 London. Published by G. Kearsley, Fleet Street, & the other Proprietors.

ing to admit this supposition; conceiving it improbable that any bird of the Parrot genus should be found at so great a distance beyond the equator; a fancy now sufficiently exploded by the numerous species of late years discovered in the southern extra-tropical regions.

TABUAN PARRAKEET.

Psittacus Tabuensis. *P. viridis, fascia humerali pallidiore, capite collo corporeque subtus coccineis.*

Green Parrakeet, with paler shoulder-stripe, and scarlet head, neck, and under-parts.

Psittacus Tabuensis β. *P. macr. viridis, capite collo corporeque subtus coccineis, &c. Lath. ind. orn.*

La Grande Perruche à collier et croupion bleu. *Levaill. pl.* 55.

Psittacus Tabuensis. *Museum Leverianum. No.* 6. *p.* 29.

Tabuan Parrot. *White's Bot. Bay. pl. p.* 168.

VAR.?

Psittacus Amboinensis. *P. macrourus coccineus, dorso cæruleo, alis macula viridi. Lin. Syst. Nat.*

Psittacus Amboinensis. *Lath. ind. orn.*

Le Lori Perruche tricolor. *Buff. ois.*

Perruche rouge d'Amboine. *Pl. Enl.* 240.

Amboina Red Parrot. *Lath. syn.*

LENGTH about sixteen, or even eighteen inches: head, neck, and whole under parts bright scarlet: back, wings, and tail deep grass-green, with an oblique bar of much brighter and paler green across the smaller wing-coverts, passing from the shoulders towards the back: larger quill-feathers tipped with dusky blackish-blue: tail dusky, with a varying cast of green: upper mandible orange-red with dusky tip; lower mandible blackish: legs of the same colour. Native of New Holland, and, in all probability, of some of the East-Indian islands, since there can be little doubt that the Psittacus Amboinensis of Linnæus is in reality the same species.

The Tabuan Parrakeet is a bird of great beauty, and has often been brought from New Holland to this country in a living state. It appears to be of a gentle, but not remarkably lively disposition.

The figure of this species in Mr. White's Journal of a Voyage to New South-Wales seems to represent the tail as nearly even at the end; a circumstance which must have arisen either from some slight accidental mutilation of the specimen, or inattention on the part of the artist; the tail, in all the specimens brought over in a perfect state, being long, and moderately cuneiform, as in the immediately preceding species.

POMPADOUR PARRAKEET.

Psittacus atropurpureus. *P. atropurpureus, dorso alis caudaque saturate holoserico-viridibus, remigibus primoribus rectricibusque exterioribus cyaneis. Museum Leverianum. p.* 140.

Dark-crimson Parrakeet, with velvet-green back, wings, and tail; the larger quill-feathers and exterior tail-feathers blue.

Pompadour Parrot. *Mus. Lev. pl. p.* 142.

Psittacus Tabuensis. *P. macr. viridis, capite collo corporeque subtus purpureo-coccineis, cervice basi lunula cærulea, remigibus rectricibusque cæruleis. Lath. ind. orn.*

Tabuan Parrot. *Lath. syn.*

General length about eighteen inches: head, neck, and whole under parts fine but very deep crimson or pompadour-colour: back, shoulders, scapulars, and tail, deep grass-green, with a rich silky or velvet-like surface; the two middle tail-feathers slightly clouded with a varying cast of blue; the two or three exterior ones inclining strongly to this colour: larger wing-feathers deep blue: bill and legs dusky or blackish. Native of New Holland.

This bird is the Tabuan Parrot of Dr. Latham, who considers the brilliant species before described as a variety of the present. The difference of colour however, and in some degree, of shape, this being of a somewhat more slender habit than the preceding, seem to indicate a specific difference, unless the darker colours of the present bird should be merely owing to a less advanced state of age and plumage.

PENNANTIAN PARRAKEET.

Psittacus Pennantii. *P. coccineus, gula alis caudaque cæruleis, dorso nigro variato.*

Scarlet Parrakeet, with blue throat, wings, and tail, and back varied with black.

Psittacus Pennantii. *P. macr. coccineus, dorso antico nigro coccineo undulato, lateribus corporis gulæque cæruleis, remigibus intus macula alba. Lath. ind. orn.*

Psittacus gloriosus. } *Nat. Misc. pl.* 53.
Splendid Parrot. }

Pennantian Parrot. *White's Journ. pl. p.* 174. *Lath. syn. suppl.*

Beautiful Lory. *Lath. syn.*

La Perruche à large queue. *Levaill. pl.* 78. 79. 80.

A SPECIES of great elegance, both in shape and plumage: length about fifteen or sixteen inches: bill horn-colour: head, neck, back, rump, and whole under parts from the throat, fine scarlet, every feather on the back being black in the middle, so as to appear deeply edged with scarlet: the scapular-feathers are also of similar appearance: throat fine blue: wings and tail the same, with a much paler and brighter gloss on the coverts or shoulder-parts: inner bend of the wing on each side the back black: tips of the larger quill-feathers black or dusky: side-feathers of the tail pale or whitish towards the tips; the two middle feathers deep blue: legs black. Native of New Holland. The female is said to be olive-green on the back, spotted in the same manner as in the male, and the tail dark blue, edged or fringed with chesnut colour.

M. Griffith sculp

PENNANTIAN PARRAKEET.

1811. Aug. 1. London, Published by G. Kearsley, Fleet Street.

NONPAREIL PARRAKEET.

1811 Aug.t 1. London Published by G. Kearsley, Fleet Street.

The Parrakeet described in Dr. Latham's Ornithology under the title of *Beautiful Lory* is certainly no other than the present species, and indeed Dr. Latham, in his second Supplement, has made the same observation. The specimen described under that name by Dr. Latham seems not to have arrived at its full colours; the red on the back being slightly tinged with green, and the black itself of a brownish cast.

NONPAREIL PARRAKEET.

Psittacus eximius. *P. capite collo pectoreque coccineis, alis caudaque cæruleis, dorso nigro viridi undulato, abdomine flavescente.*

Parrakeet, with scarlet head, neck, and breast, blue wings and tail, black back undulated with green, and yellowish abdomen.

Perruche omnicolore. *Levaill.* 28. 29.

Psittacus eximius. *P. macrourus varius, capite gula pectore crissoque coccineis, dorso nigro flavo-viridi undulato, alis caudaque cæruleis. Vivar. Nat. t.* 93.

Nonpareil Parrot. Long-tailed variegated Parrot, with head, throat, breast, and vent crimson, back black undulated with yellow-green, and blue wings and tail. *Nat. Misc. pl.* 93. *New Holl. Zool. pl.* 1.

Psittacus eximius. } *Lath. suppl.* 2.
Nonpareil Parrot. }

THIS most beautiful species, which was first described by myself in the Naturalist's Miscellany, and afterwards in the Zoology of New Holland, is in size, shape, and general distribution of colours, greatly allied to the Pennantian Parrakeet: the colours themselves however, on some parts of the bird, are widely different: the bill is horn-coloured; the upper mandible palest: the head, neck, breast, and upper part of the belly of a beautiful scarlet: the throat white with a slight tinge of blossom-colour: the wings fine mazarine blue, with a paler glossy lustre on the middle of the covert-feathers: the prime quills dusky blue with pale edges: the smaller coverts next the back black: hind part of the neck, scapulars, and whole back, as far as the rump, black, each feather elegantly bordered with bright apple-green: lower part of the belly of the same colour, but yellower, and in some specimens slightly undulated with pale red: tail fine blue, growing pale or whitish toward the tips; but the two middle feathers blue-green: legs cinereous. Native of New Holland, from whence it has been often brought into England. In a state of captivity it appears of a mild disposition, but has not, I believe, been observed to articulate. It is one of the most beautiful Parrakeets yet discovered.

M. Griffith sculp.

NONPAREIL PARRAKEET *var.*

1811. Aug.t 1. London, Published by G. Kearsley, Fleet Street, & the other Proprietors.

BLUE-BELLIED PARRAKEET.

1811, Aug.st 1. London, Published by G. Kearsley, Fleet Street, & the other Proprietors.

BLUE-BELLIED PARRAKEET.

Psittacus cyanogaster. *P. viridis capite cæruleo, subtus luteo rubro cæruleoque varius, remigibus intus macula lutea.*

Green Parrakeet, with blue head, beneath varied with yellow, red, and blue; the quill-feathers marked within by a yellow spot.

Psittacus hæmatodus var. β. γ. δ. *Lath. ind. orn.*

Red-Breasted Parrot var. A. B. *Lath. syn.*

Perruche des Moluques. *Buff. ois. Pl. Enl.* 743.

Perruche d'Amboine. *Buff. ois. Pl. Enl.* 61.

Blue-Bellied Parrot. *White's Journ. pl. p.* 140.

Perruche à tete bleue. *Levaill.* 24. 25. 26. 27.

THIS is a very beautiful and richly variegated species, and appears to occur, with some variations, through a large extent of space; being found in several of the Molucca islands, as well as in New Holland, where it occurs in the highest state of perfection, generally measuring about fourteen or fifteen inches in length: the bill is pale orange-red: the head and throat fine violet-blue, the feathers on those parts being somewhat lanceolate: the neck, back, and wings deep grass-green, but the larger quill-feathers dusky, and crossed on the inner webs by a deep-yellow bar: the breast deep red, waved or mottled transversly with bright yellow: the under coverts of the wings, beneath the shoulders, are red: the belly fine blue, and the thighs varied with green and yellow; the legs dusky. The tongue in this, as well as in some other New Holland Parrakeets, is finely ciliated, as it were, on each side the tip, by a number of lengthened white papillæ or processes. The Blue-Bellied Parrakeet is a bird of a lively disposition, and may be rendered perfectly familiar and caressing. It feeds readily, in a state of captivity, on bread soaked in milk or water, as well as on various kinds of fruits.

Monsr. Levaillant, during his residence at the Cape of Good Hope, had an opportunity of contemplating a pair of this species, which were imported from Amboina. These birds bred during their confinement in the menagerie of Monsr. Van Bletemberg, then governor at the Cape. The female deplumed her breast, and after having collected the feathers into a heap, deposited two round white eggs, on which she sat most assiduously; the male feeding her at intervals, by disgorging what he had swallowed into his beak, and presenting it to her. The young were produced at the end of nineteen days, and in the space of a few more were covered with a cinereous-grey down, which was by degrees succeeded by green feathers on the body, by blue ones on the head. At the end of three weeks they left the nest, and perched on the neighbouring sticks, where the male and female in concert fed them as above described, in the manner of pigeons. The parent birds continued to feed them in this manner for six months, and often afforded a very interesting scene; the young being frequently seated beyond the female; and the male, not being able to reach them, first presented the food to the female, who immediately delivered it to her young in succession. The young, though of different sexes, were perfectly alike till the first moulting, at which time red feathers, bordered with green, began to appear on the breast, and the male became distinguished by the blue patch on the abdomen. This species is most elegantly figured, in its several states, in the splendid work of Monsr. Levaillant, who has also given a plate of a remarkable variety, in which the back and coverts are yellow, and the sides both of the neck and lower part of the abdomen richly scalloped with the same colour.

VAR.?

RED-BREASTED PARRAKEET.

Psittacus hæmatodes. *P. macr. viridis, pectore rubro, facie cærulea, lunula cervicis flava. Lin. Syst. Nat. Mantiss.* 1771. *p.* 524.
Red-Breasted Parrakeet. *Edw. pl.* 232.
La Perruche à face bleue. *Buff. ois.*
Perruche d'Amboine. *Pl. Enl.* 61.
La Perruche a face bleue. *Levaill. pl.* 47.

Monsr. Levaillant considers this, which was first described by Edwards, under the name of the Red-Breasted Parrakeet, as a distinct species, having had an opportunity of viewing three specimens in different collections, all which exactly coincided in colours with Edwards's figure and description, and were all said to be natives of some part of the East Indies, though their precise native region seemed unknown. Levaillant observes that, exclusive of its colours, it differs in having the tail-feathers much more pointed than those of the Blue-Bellied Parrakeet.

VARIEGATED PARRAKEET.

Psittacus ornatus. *P. viridis luteo variatus, vertice cæruleo, occipite gula pectoreque rubris atro-viridi undulatis.*

Green Parrakeet varied with yellow, with blue crown, and red hind-head, throat and breast undulated with blackish-green.

Psittacus ornatus. *P. macrourus luteo-viridis, occipite gula pectoreque rubris, vertice auribusque cæruleis, orbitis cinereis. Lin. Syst. Nat.*

Lory Parrakeet. *Edw. pl.* 174. *Lath. syn.*

La Perruche Lori. *Buff. ois.*

La Perruche variée des Indes. *Pl. Enl.* 552.

In its general appearance, and disposition of colours, this species is strongly allied to the Psittacus cyanogaster or Blue-Bellied Parrakeet, but is of smaller size, with a much shorter tail in proportion: the general length of the bird is about eight inches: the bill is orange-coloured; the fore-part of the crown violet-blue; the hind-part crimson with the feathers edged with violet; the orbits of the eyes bare and of a blue-grey colour: from behind each eye springs a short broad violet stripe, pointing backwards, and immediately beyond this is an orange yellow stripe: the chin, throat, and breast are crimson, the feathers edged either with dark green or violet-colour: the remainder of the

VARIEGATED PARRAKEET.

1811 Augt 1. London Published by G. Kearsley, Fleet Street & the other Proprietors.

bird is grass-green, scalloped or varied on the upper part of the back, and sides of the body with orange-yellow: the tail is green above, but orange-red beneath, with yellowish-green tips: the legs deep ash-colour.

So great is the general similarity between this bird and the Blue-Bellied or Red-Breasted Parrakeet, that on a cursory view it might pass for a small specimen of the same species, with a shorter tail than ordinary.

JAPONESE PARRAKEET.

Psittacus Japonicus. *P. macr. viridis, subtus rectricibusque lateralibus ruber, remigibus cæruleis. Lin. Syst. Nat.*
Psittacus erythrochlorus macrourus. *Aldrov.*
La Perruche verte et rouge. *Buff. ois.*
Japonese Parrot. *Lath. syn.*

Slightly described by Willughby, Linnæus, Buffon, and others from Aldrovandus, whose description rests merely on the faith of a Japonese drawing; so that the very existence of the bird is by no means fully ascertained. It is to be numbered, according to Aldrovandus, among the middle-sized Parrots, being about the same size with the common Ring-Parrakeet: the bill is very small, and different from that of other Parrots in being curved beneath: the whole b[illegible] crown of the head, neck, and wing-coverts are deep green, except at the scapulæ, where the upper feathers are blue: all the quill-feathers are deep blue, except the two exterior ones, but all of them have white shafts: the irides are red: the bill red, and very slightly curved: the chin ferruginous or reddish: before and behind the eyes is a blue spot: the whole breast and belly are of a vermilion red, the tips of the feathers being marked with small longitudinal lines: the tail is very long in proportion, and even longer than the whole body; the two larger feathers are greenish with white shafts; the interior ones are deep red with black shafts; the legs and feet deep black.

RED-TOPPED PARRAKEET.

Psittacus verticalis. *P. macrourus viridis, vertice medio rubro, remigibus cœruleis. Lath. ind. orn. suppl.*

Green Parrakeet, with the middle of the crown red, and blue quill-feathers.

Red-topped Parrot. *Lath. suppl.* 2.

LENGTH eighteen inches: bill large, blue, with a black tip: colour of the bird dark green, paler beneath: legs brown. Native of New Holland: described by Dr. Latham, who observes, that it is most allied to the Pacific Parrakeet, but seems to be nearly twice the size.

VAR. ?

CRIMSON-CROWNED PARRAKEET.

Size of Tabuan Parrakeet, to which it seems extremely allied: colour similar, except that the crown of the head alone is of a rich scarlet: across the shoulders a pale-green bar, as in the Tabuan: rump blue: colour of the under parts uncertain, the bird having been described from a specimen in a glass case, and so disposed as to exhibit the upper parts only. Supposed to be a native of New Holland.

PACIFIC PARRAKEET.

Psittacus Pacificus. *P. macr. viridis, fronte temporibus uropygioque lateribus rubris. Lath. ind. orn.*

Green Parrakeet, with the forehead, temples, and sides of the rump scarlet.

Pacific Parrot. *Lath. syn.*

Psittacus australis. *Lath. ind. orn.*

Psittacus concinnus. *Nat. Misc. pl. 87.*

Size of a small Turtle: general length about twelve inches: colour deep grass-green, paler beneath: behind each eye a short, pointed spot or stripe of the same colour: crown pale blue: back of neck and tips of wings pale brown: tail of very moderate length, but strongly cuneated and sharp-pointed: bill blueish; legs dusky. In some specimens a red spot appears on each side the vent, and in some the crown of the head is green like the rest of the plumage; in others yellowish, and the rump red. Native of New Holland and several of the southern isles.

VARIED-WINGED PARRAKEET.

Psittacus marginatus. *P. viridis, vertice cæruleo, tectricibus alarum nigro cæruleo flavoque variis.*

Green Parrakeet, with blue crown, and wing-coverts varied with black, blue, and yellow.

La Perruche de l'isle de Luçon. *Sonner. voy. p.* 80. *t.* 44.

Varied-Winged Parrot. *Lath. syn.*

Psittacus marginatus. *Lath. ind. orn.*

Described by Sonnerat, who says it is considerably larger than the common Parrakeet, but does not mention its length: bill large and flesh-coloured: eyes small: irides white: crown of the head blue: upper parts of the bird grass-green; under parts yellowish-green: smaller wing-coverts black, edged with yellow-brown; greater coverts black, edged with blue, and again with yellow-brown, so as to form a richly variegated patch on the wings: legs blackish. Native of Luzonia.

VAR. ?

Psittacus phrygius. *P. viridis, vertice cæruleo, tectricibus alarum cæruleis flavo marginatis.*

Green Parrakeet, with blue crown, and blue wing-coverts edged with yellow.

Psittacus olivaceus. *Lath. ind. orn. Lin. Syst. Nat. Gmel.*

Perruche à ailes chamarreès. *Buff. ois.*

Perroquet de l'isle de Luçon. *Pl. Enl.* 287.

Lace-winged Parrakeet. *Lath. syn.*

La Perruche aux ailes chamarreès. *Levaill. pl.* 60.

THERE is great reason for supposing this to be a mere variety, owing to a different stage of plumage; or a sexual difference of the preceding. Its total length, according to Levaillant, is from twelve to thirteen inches; and in size of the body it is nearly equal to a common Grey Parrot: its colour is deep green, somewhat yellower beneath: the crown of the head is blue, the colour forming a broad transverse bar on the head: the wing-coverts are blue, with yellow borders, and the alulet or spurious wing green: bordered in a similar manner: the larger quill-feathers dusky, with yellow edges: the bill large and red; the tail of very moderate length in proportion to the bird, and not very sharply cuneated. In the Planches Enluminées this bird is represented of an olive-colour, and this accounts for Buffon's erroneous description of the colours, since he may be supposed, according to Monsr. Levaillant, to have made up his description from the figure in that publication.

Levaillant also adds, that the name of *Perroquet de Luçon*, by which it is called in the work just mentioned, is highly exceptionable, since it is found not only through a great extent of the East Indies, but in all the Molucca islands.

RED-RUMPED PARRAKEET.

Psittacus Zælandicus. *P. macr. fusco-viridis, capistro purpureo-nigro, vertice viridi-castaneo, striga per oculos uropygioque coccineis. Lath. ind. orn.*

Dusky-green Parrakeet, with blackish-purple frontlet, greenish-chesnut crown, red eye-stripe and rump.

Red-rumped Parrot. *Lath. syn.*

FIRST described by Dr. Latham, from a specimen then in the possession of Sir Joseph Banks. Length fifteen inches: colour above dusky green, beneath cinereous green: quill-feathers with blueish edges: forehead dark purple: crown greenish chesnut: from the base of the bill, through the eyes, a crimson streak: rump crimson with a tinge of chesnut: tail blueish, the two middle-feathers edged with green, and the shafts of all the feathers deep chesnut: bill stout, and dusky-blue at the base: legs black. Native of New Zealand.

ALEXANDRINE PARRAKEET.

London Published by G. Kearsley, Fleet Street, & the other Proprietors.

ALEXANDRINE PARRAKEET.

Psittacus Alexandri. *P. viridis, striga humerali collarique postico rubris, gula collarique antico nigris.*

Green Parrakeet, with red hind-collar and shoulder-stripe, and black fore-collar and throat.

Psittacus torquatus macrourus atiquorum. *Aldr. orn. Will. orn.*

Psittacus Alexandri. *Lin. Lath. &c. Pl. Enl.* 642. *Edw. pl.* 292. *Levaill. pl.* 30.

This elegant species, which has so long maintained a distinguished reputation for its docility and imitative powers, is supposed to have been the only bird of the Parrot kind known to the ancient Greeks and Romans, having been brought from the island of Ceylon, (the ancient Tabropane,) after the Indian expeditions of Alexander the Great. In the reign of Nero the Romans first became acquainted with other species of Parrots, which they obtained from various parts of Africa. The description given of the present species by Pliny, however short, was sufficient to distinguish it from all others at the time he wrote. It is, says he, entirely green, with a red collar on the neck. These birds appear to have been in great request among the Romans, who lodged them in superb cages, ornamented with silver, tortoise-shell, and ivory; and the price of a Parrot often exceeded that of a slave.

The size of the Alexandrine or Ring Parrakeet is that of a common Pigeon, its general length about fifteen inches, and its colour an elegant bright green above, paler or yellower beneath: across each shoulder, on the smaller coverts, is a lengthened purplish-red patch or spot, and from the base of the lower mandible, on each side, proceeds a moderately broad black band or stripe, which after descending a little way, passes backwards, so as almost to encircle the neck, growing very narrow as it approaches the back part, which is marked by a red collar, near half an inch wide, but narrowing as it passes forwards immediately beneath the black one, almost reaching the front of the neck: the back part of the head, towards the commencement of the red collar, has a slight blueish tinge, and the edges of the tail-feathers are often of a similar cast: the bill is of a bright orange-red; the legs ash-coloured; and the under surface of the tail, which is strongly and regularly cuneated, is of a yellowish cast.

A variety of this bird is described and figured by Edwards, which in size and all other particulars resembled the common kind, except in being the head of a sky-blue colour, without any appearance of a collar round the neck; the tail pale blue above, and the spot on the shoulders dull yellow instead of red. Exclusive of this, some other kinds appear to exist, retaining the leading characteristics of the species, but, like the race of pigeons, constituting so many permanent varieties. Of these the following are enumerated by the Count de Buffon, Dr. Latham, and others.

ROSE-RINGED PARRAKEET.

Size of a Blackbird: length rather more than fifteen inches: upper mandible blood-red with a black tip; lower black: eyes surrounded by a flesh-coloured skin: colour of the plumage pale green, the hind-head inclining to violet: round the neck a ring of light rose-colour, and which is not visible till the third year of the animal's growth. This variety is a native of several parts of Africa. Monsr. Levaillant considers it as a distinct species.

DOUBLE-RINGED PARRAKEET.

Described by Brisson and Buffon: length thirteen inches and a half: bill as in the preceding kinds: plumage deep green, yellower beneath: the ring round the neck rose-colour, and broadest in front, while from each side of the lower mandible passes a narrow black stripe, accompanying the red bar as it passes backwards: the plumage of the neck, immediately above the collar, being tinctured with blue-green, and beneath it with yellowish. Native of the East Indies and some parts of Africa.

PURPLE-RINGED PARRAKEET.

LENGTH seventeen inches: bill as in the immediately preceding: collar round the neck purple, and the fore part of the neck and breast pale rose-colour. This is figured in the work of Albin, and is said to come from the East-Indies. It is perhaps rather the Mustachoe Parrakeet ill figured.

BLUE-COLLARED PARRAKEET.

SIZE of a common Parrakeet: bill and irides red: head, neck, and belly greyish-green: on the neck a bar of pale lilac-colour, forming a collar: wings and tail grass-green: on each shoulder a large deep red spot. Native of the island of Luzonia, where it was observed by Sonnerat, and though sometimes considered as a distinct species, seems too nearly allied to the Alexandrine to be regarded in any other light than a variety.

JAVAN PARRAKEET.

LESS than the common Ring-Parrakeet: upper mandible pale red; lower pale yellow: head of a mixed pale blue and pale yellow: temples black: remainder of the bird grass-green, except that the thoat and breast are pale red, and five of the wing-coverts yellow: tail yellowish: legs greenish-grey. This is described by Osbeck, who observed it in Java, and considers it as a variety of the Alexandrine.

BLUE-HEADED PARRAKEET.

Described by Brisson, who says it is of the size of a common Parrakeet, and about eleven inches and a half in length: the upper mandible yellow with a cinereous tip, the lower cinereous: head blue with a tinge of red on the forehead, and of violaceo-cinereous on the throat: colour of the upper parts deep shining green, the wings tipped with cinereous, the shafts of the quill-feathers black: upper part of the breast, and under parts of the bird tinged, or rather varied with yellow: tail-feathers blueish towards the tips, and yellowish on their inner webs: legs ash-coloured. Native of the East Indies. Quer. Whether it be any thing more than the blue-headed variety already described, with the additional circumstance of being varied beneath with yellow?

JONQUIL PARRAKEET.

This most elegant bird is described and figured as a distinct species in Dr. Latham's second Supplement, from a drawing communicated by General Davies, taken from the living bird, which was said to have come from the province of Bahar in Bengal. Its length is about ten inches and a half; its habit that of the Ring-Parrakeet, and the tail very long and slender in proportion: the colour of the plumage is a fine jonquil yellow, paler beneath, but the head of a bright crimson, bounded on the back of the neck by a sea-green collar, narrowing as it approaches forwards, where it meets a white one arising from the sides of the lower mandible: on the shoulders, at the bend of the wing, is an oblong red patch, and the bill and legs are flesh-coloured.

SULPHUR PARRAKEET.

LENGTH about fifteen inches: habit that of the Alexandrine Parrakeet: colour uniform pale or sulphur-yellow, rather deeper on the back: bill, legs, and feet pale: described and figured by Levaillant, from a preserved specimen in a collection at Leyden: uncertain whether a distinct species, or a variety of some other.

Monsr. Levaillant reasons well on the subject of the varieties with respect to plumage which so often take place in the Parrot tribe. All birds in general, he observes, are subject to become white, as we know from the numerous examples daily before our eyes; even such birds as are naturally of the opposite colour, as Ravens, Magpies, Blackbirds, &c. There are also white varieties of Thrushes, Jays, Partridges, Snipes, and Woodcocks, Sparrows, Swallows, Martins, and Goatsuckers. It has been imagined that such changes were owing to age; but, on the contrary, it is certain that these variations from the genuine colour are always observed to take place in young birds or nestlings, and these birds at their first moulting sometimes recover, either wholly or in part, their proper colours. Neither is this change confined, as is often supposed, to the birds of Northern Climates, but takes place equally in those of Africa and South-America. There is however, adds Monsr. Levaillant, no example of any kind of the Parrot tribe becoming white or varied with white, (exclusive, of course, of the Cockatoos,) which are naturally of that colour. We frequently however observe several of the Parrot tribe to become patched with yellow, and even to become entirely of that colour, however different their natural plumage may have been. We find that throughout Nature yellow forms the base of green, which is the prevailing colour in the Parrot tribe. Thus the leaves of trees, when fading, or dried, turn yellow. This colour also, according to Monsr. Levaillant, is the basis of red; and from the whole he concludes that yellow is to the Parrot tribe what white is to the generality of birds.

To return to the individual above described, or Sulphur Parrakeet, Monsr. Levaillant considers it as most allied to the Rose-Ringed Parrakeet, of which he therefore supposes it may be a variety; though he does not presume absolutely to pronounce it such.

It would be unpardonable to dismiss the history of the Alexandrine or Ring-Parrakeet without observing that it was in commemoration of a bird of this species that Ovid's celebrated elegy on the death of Corinna's Parrot was composed, and which may justly be considered as one of the most elegant trifles that antiquity can boast. In the full confidence that it cannot be unacceptable to the poetical and sentimental reader, I here subjoin it, accompanied by a free translation.

Psittacus, Eois imitatrix ales ab oris,
  Occidit. exequias ite frequenter aves.
Ite, piæ volucres, et plangite pectora pennis;
  Et rigido teneras ungue notate genas.
Horrida pro moestis lanietur pluma capillis:
  Pro longa resonent carmina vestra tuba.
Quid scelus Ismarii quereris Philomela tyranni?
  Expleta est annis ista querela suis.
Alitis in raræ miserum devertite funus.
  Magna, sed antiqui causa doloris Itys.
Omnes quæ liquido libratis in aëre cursus
  Tu tamen ante alias turtur amice dole.
Plena fuit vobis omni concordia vita,
  Et stetit ad finem longa tenaxque fides.
Quod fuit Argolico juvenis Phocæus Orestæ,
  Hoc tibi, dum licuit, Psittace, turtur erat.
Quid tamen ista fides? quid rari forma coloris?
  Quid vox mutandis ingeniosa sonis?
Quid juvat, ut datus es, nostræ placuisse puellæ?
  Infelix avium gloria, nempe jaces.
Tu poteras virides pennis hebetare smaragdos,
  Tincta gerens rubro punica rostra croco.

Non fuit in terris vocum simulantior ales;
  Reddebas blæso tam bene verba sono.
Raptus es invidia. Non tu fera bella movebas:
  Garrulus, et placidæ pacis amator eras.
Ecce, coturnices inter sua prælia vivunt:
  Forsitan et fiant inde frequenter anus.
Plenus eras minimo; nec præ sermonis amore
  In multos poteras ora vacare cibos.
Nux erat esca tibi, causæque papavera somni;
  Pellebatque sitim simplicis humor aquæ.
Vivit edax vultur, ducensque per aëra gyros
  Miluus, et pluviæ graculus auctor aquæ.
Vivit et armiferæ cornix invisa Minervæ;
  Illa quidem sæclis vix moritura novem.
Occidit ille loquax, humanæ vocis imago,
  Psittacus, extremo munus ab orbe datum.
Optima prima fere manibus rapiuntur avaris;
  Implentur numeris deteriora suis.
Tristia Phyllacidæ Thersites funera vidit:
  Jamque cinis, vivis fratribus, Hector erat.
Quid referam timidæ pro te pia vota puellæ;
  Vota procelloso per mare rapta Noto?
Septima lux aderat, non exhibitura sequentem,
  Et stabat vacua jam tibi Parca colo.
Nec tamen ignavo stupuerunt verba palato:
  Clamavit moriens lingua, Corinna, vale.
Colle sub Elysio nigra nemus ilice frondens,
  Udaque perpetuo gramine terra, viret.
Si qua fides dubiis; volucrum locus ille piarum
  Dicitur, obscænæ quo prohibentur aves.
Illic innocui late pascuntur olores:
  Et vivax phoenix, unica semper avis.
Explicat ipsa suas ales Junonia pennas:
  Oscula dat cupido blanda columba mari.
Psittacus has inter, memorali sede receptus,
  Convertit volucres in sua verba pias.
Ossa tegit tumulus; tumulus pro corpore parvus:
  Quo lapis exiguus par sibi carmen habet.
Colligor ex ipso dominæ placuisse sepulchro.
  Ora fuere mihi plus ave docta loqui.

Born in the fav'ring blaze of Eastern skies,
Corinna's pride, the feather'd talker dies.
Attend, ye tuneful denizens of air,
And join your sorrows with the weeping fair.
With flagging wings and ruffled plumes appear,
And sing a mournful requiem round his bier.
Sad Philomel, thy notes no longer ply
On Ity's fate and Tereus' cruelty.
If still thou call'st it pleasure to complain,
Now let the Indian favourite claim thy strain.
But chief, mild Turtle, o'er his body bend,
And wail in lengthen'd notes the parted friend.
With thee thro' life his fond affection ran;
Nor death extinguish'd what so well began.
Nor Pylades with his Orestes join'd
E'er found a friend more constant or more kind.
Now what avails, alas! that sacred name?
Or what the beauties of thy matchless frame?
Or what the mimic voice, that well could try
Each varying word and changing minstrelsy?
Ill-fated praise! nor could thy mistress save
Her feather'd favourite from a timeless grave.
Scarce could the emerald match thy verdant quill;
While saffron tinged the ruby of thy bill.
Thy temperance no sage hath e'er surpass'd:
Nuts, poppies, water, form'd the sole repast.
Thy speech excell'd thy own loquacious race,
And lisp'd each falling word with happier grace.
Oft would thy ready tongue each accent steal,
While talk supplied thy interrupted meal.
Invidious Fate! with ruthless hand to seize
A harmless, peaceful prattler, born to please!
Yet fretful Quails amid' their contests live,
And thro' the series of their wounds survive.
Yet the fierce Vulture lives, and tears his prey;
And the Kite wheels in air his circling way.

The boding Raven mocks the shafts of fate;
And scarce nine ages fill his endless date.
The worthless Jackdaw hovers o'er the plain,
And calls with noisy throat th' impending rain.
But thou, sweet copier of the human strain,
Art gone; and India sent her bird in vain!
Fate seizes first the worthiest and the best;
And, with perverse forbearance, spares the rest.
The base Thersites Peleus' son surviv'd;
And Hector perish'd, while his brethren liv'd.
Why should I tell thy gentle mistress' prayers,
Giv'n to the winds, tho' urg'd with falling tears?
Sev'n days had now their gloomy tenor run,
And Fate, her distaff bare, thy thread had spun.
E'en then ere yet the power of speech had pass'd,
Farewell, belov'd Corinna! was thy last.
Deep in Elysian vales, in waving rows
Spread the green shades where pious birds repose:
And whence debarred, no place of rest can claim
Or fowl of cast obscene, or evil fame.
But peaceful Swans their silver wings display,
And gently glide along their watery way.
There too, 'tis said, the living Phoenix blooms,
And waves in purple skies his golden plumes.
There Juno's birds, with harmless pride, display
The varying gems that o'er their plumage play.
While, deep in myrtle shades, the cooing Doves
Breathe their soft murmurs, and indulge their loves.
There too, receiv'd amid the sacred groves,
Corinna's pride, the emerald stranger roves;
While, gathering round him, the selected quire
Catch the new accents, and their notes admire.
His bones beneath their little tomb repose:
His character this short inscription shows.
Here lies the glory of the feather'd race;
The first in language as the first in grace.

ROSE-HEADED PARRAKEET.

Psittacus rodocephalus. *P. viridis, capite roseo-cæruleo, gula et torque nigris, rectricibus mediis cæruleis apice albidis.*

Green Parrakeet, with rose-blue head, black throat and collar, and blue middle tail-feathers with whitish tips.

Psittacus rodocephalus. *Museum Leverianum. p.* 83. *Vivarium Naturæ, vol.* 21. *pl.* 877.

La Perruche Fridytutah. *Levaill. pl.* 74.

Psittacus Ginginianus. var. *β*. *Lath. ind. orn.*

A most beautiful species, of smaller size, in general, than the Alexandrine Parrakeet, but greatly allied to it in the general distribution of its colours: it differs however in the shape of the tail, the two middle feathers of which extend far beyond the rest. In length this species seems to vary from twelve to fifteen inches: its plumage is a fine green, paler or yellower beneath: on the shoulders is a small oblique purple-red patch or spot: the upper mandible is bright orange-colour, the lower black: the front and cheeks are of a fine rosy-red, which colour on the remainder of the head gradually declines, and is overcast by a strong tinge of full, yet softened blue, which, perhaps, can be explained by nothing so well as by the appearance of that beautiful bloom which shews to so much advantage on the surface of the Orleans plum. From the base of the lower mandible, round the neck, proceeds a narrow collar of the deepest black; immediately beneath which is a

ROSE HEADED PARRAKEET.

corresponding circle of sea-green, a tinge of which colour is also observable in the shoulders and rump: the two middle tail-feathers are of a fine deep ultramarine blue, with white tips; the two next pair have yellowish tips, while the remaining ones are blueish-green on their exterior sides, and yellowish on the interior: the legs and feet are pale grey. This species is a native of India and some of the Indian islands, and varies somewhat as to the colour of the head in different individuals; the deep rose-colour prevailing in some, and the blue in others, while in some the breast is tinged with rose-colour, and in others the head has a cast of green, and the ring round the neck is wanting; perhaps from not having arrived at full perfection of plumage. A most beautiful specimen of the Rose-Headed Parrakeet was preserved in the Leverian Museum, and is described and figured in the fourth number of the work entitled Museum Leverianum. This species is also most elegantly figured in the superb work of Monsr. Levaillant, under its Indian name of Fridytutah.

MUSTACHOE PARRAKEET.

Psittacus mystaceus. *P. viridis, vertice cano, fascia frontali maculaque utrinque postmandibulari nigris, pectore roseo.*

Green Parrakeet, with grey crown, black frontal bar, black spot behind each side the bill, and rose-coloured breast.

Psittacus Pondicerianus. Mustachoe Parrakeet. *Lath.*

Perruche à moustaches. *Buff. ois.*

La Perruche a poitrine rose. *Levaill. pl.* 31.

Bracelet Parrakeet? *Albin.* 2. *pl.* 18.

This species seems to have been first described by the Count de Buffon, and is a native of India. Its length varies from twelve to fifteen inches: the colour of the plumage dark green above, but the smaller wing-coverts yellow, and the larger edged with that colour: the head grey, with a narrow black band across the forehead, from eye to eye immediately above the base of the upper mandible, while from each corner of the lower mandible springs a black streak, widening, as it passes backwards, into the appearance of a rounded whisker, and giving the bird a very singular aspect when viewed in front: the throat is white, the breast and belly pale rose or blossom-colour, and the larger wing-feathers and tail blueish; the tail straw-coloured beneath, the two middle feathers considerably exceeding the rest in length. A bird of this species which I examined, appeared to be of a gentle disposition, but by no means lively, and was remarkable for being best pleased when placed

MUSTACHOE PARRAKEET.

1811, Aug. 1. London, Published by G. Kearsley, Fleet Street, & the other Proprietors.

in an obsure corner of the room in which it was kept. From the singular appearance of this species when viewed in front, it has obtained among the bird-dealers the name of the Jew Parrakeet.

BLOSSOM-CHEEKED PARRAKEET.

Psittacus erubescens. *P. viridis, genis nuchaque roseo-violaceis, macula utrinque postmandibulari nigra.*

Green Parrakeet, with rose-violet cheeks and nape, and a black spot behind each side the bill.

Grande Perruche à longs brins. *Buff. ois.*

Perruche de Malac. *Pl. Enl.* 887.

Psittacus Ginginianus. Blossom-Headed Parrakeet. var. C. *Lath.*

Perruche à nuque et joues rouge. *Levaill. pl.* 72.

Length from sixteen to eighteen inches: general habit that of P. Alexandri, but with an arrow-shaped tail, or with the two middle feathers much exceeding the rest in length: colour fine green: the back and shorter quill-feathers tinged with blue, and the two middle tail-feathers with violet: cheeks and nape blueish rose-colour, and on each side the base of the lower mandible a broad black whisker-shaped spot, as in the immediately preceding bird: upper mandible red, lower dusky: wings glossy-black beneath; tail dull yellow: legs and feet dusky.

GINGI PARRAKEET.

Psittacus Eupatria. *P. viridis, subtus subflavescens, rostro fasciaque humerali rubris, rectricibus mediis attenuato-elongatis.*

Green Parrakeet, yellowish-green beneath, with red bill and shoulder-stripe, and lengthened narrow middle tail-feathers.

Psittacus Eupatria. *Lin. Lath.*

Perruche de Gingi. *Pl. Enl.* 239.

Perruche à Epaulettes rouge. *Levaill. pl.* 73.

Length twenty inches: whole bird above deep grass-green, except a lengthened red spot towards the bend of the wing, and the two middle tail-feathers, which are slightly tinged with blue: under parts of the bird pale green with a yellowish mixture: bill and legs red: the tail in this species is very long, and sharply cuneated or rather arrow-shaped, the two middle feathers much exceeding the rest in length: said to be a native of Gingi in the East-Indies. Dr. Latham observes that the eyes are surrounded by a naked reddish skin, and this circumstance forms a part of the Linnæan specific character of the bird. Monsr. Levaillant however does not mention this particular, nor does any such appearance occur in his elegant plate of this species.

YELLOW-COLLARED PARRAKEET.

Psittacus flavitorquis. *P. viridis, subtus subflavescens, capite violaceo, collari flavo, rectricibus mediis cæruleis apice albidis.*
Green Parrakeet, yellowish-green beneath, with violet head, yellow collar, and blue middle tail-feathers with whitish tips.
Perruche à collier jaune. *Levaill. pl.* 75. 76.

An elegant species: general shape or habit similar to that of P. rodocephalus, but a size smaller: head violet-blue, but tinged in front with brown: neck surrounded by a jonquil-yellow collar; remainder of the plumage green, but somewhat paler or yellower beneath: largest or lowest wing-coverts tinged with blue, and the edges of the larger quill-feathers with pale yellow: tail as in the rodocephalus, viz. green, with the two middle feathers sky-blue, pretty deeply tipped with yellowish-white: bill yellowish: legs dusky. Female like the male, but with pale-violet head, without any appearance of brown in front, and the yellow collar less conspicuous. Native of India, and considered by Levaillant as a species before undescribed.

PAPUAN PARRAKEET.

Psittacus Papuensis. *P. coccineus, alis caudaque viridibus, nucha atrocæruleo, lateribus rectricibusque luteo variis.*

Scarlet Parrakeet, with green wings and tail, the hind-head varied with deep-blue, the sides of the body and tail with deep yellow.

Psittacus Papuensis. Papuan Lory. *Lath.*

Perruche Lori-Papou. *Levaill. pl. 77.*

DESCRIBED by Sonnerat: length sixteen inches: bill red: head, neck, breast, and belly carmine-colour: on the back of the head a blue spot, beyond which are two black crescents: wings grass-green, the colour extending across the middle of the back: lower part of the back red, with a blue stripe down the middle: on each side the wing, at the base is an oblong yellow spot; and above each thigh a large yellow spot: tail longer than the rest of the bird, and green for two-thirds, the remainder being yellow: legs red. Native of Papua or New Guinea, where two or three varieties of this species are said to have been observed; in one of which a broad black bar, edged with green on its upper part, passed across the belly; and in another the tail was shorter than described by Sonnerat; the rump blue; the hind part of the neck blackish-blue; the thighs yellow, and the breast marked by a yellow spot: in others the blue-black marks on the hind part of the head were wanting; the middle of the belly was green; and the

PAPUAN PARRAKEET.

1811, Aug.t 1. London, Published by G. Kearsley, Fleet Street, & the other Proprietors.

yellow on the sides came forwards so as almost to form a crescent across the breast. It should be added that the tail in this beautiful species is most strongly arrow-shaped or pointed, the two middle feathers extending far beyond the rest.

SINCIALO PARRAKEET.

Psittacus Sincialo. *P. viridis, humeris subflaventibus, cauda longa apicibus cæruleis, mandibula superiore orbitis pedibusque incarnatis.*

Green Parrakeet, with subflavescent shoulders, long tail with blue tips, and flesh-coloured upper mandible, orbits, and legs.

Psittacus rufirostris. *Lin.*

Long-tailed green Parrakeet. *Edw. pl.* 175. *Lath.*

Le Sincialo. *Buff. ois. Levaill. pl.* 42.

This, says Aldrovandus, is of equal length with the Ring Parrakeet, but smaller bodied, being not bigger than a Thrush: the bill is red, especially on the upper part, the edges and tip being blackish: the irides red or saffron-coloured: the rest of the bird of a pleasant grass-green colour, but the belly paler, and the quill-feathers deeper: the tail narrow, ending almost in a point, and near nine inches in length: the feet and legs differ from other Parrots in colour, being reddish or flesh-coloured: it is found in the American island Hispaniola, (St. Domingo,) where it is called *Scincialo.* The Italians, adds Aldrovandus, from its small size, name it Perochino, and the French Perroquet.

According to Dr. Latham the length of this bird

is twelve inches and a quarter; the bill blood-red, and black at the point; the under mandible wholly black; the eyes situated in a bare flesh-coloured skin.

It is considered as a lively and docile bird; is very noisy in its natural state, and when tame is easily taught to articulate, and shews considerable powers of imitation. It is found not only in St. Domingo, but in various parts of South-America.

VAR.

In Brasil it varies with the bill entirely flesh-colour; and it is to this variety that the specimen figured by Edwards is to be referred. Edwards observes that the tail-feathers are nearly blue towards their tips, which are very short.

JAQUILMA PARRAKEET.

Psittacus Jaquilma. *P. macr. viridis, remigibus apice fuscis, orbitis fulvis. Lath. ind. orn.*

Green Parrakeet, with fulvous orbits, and quill-feathers dusky at the tips.

Psittacus Jaquilma. *Molina Chil. p.* 228.

Jaquilma Parrot. *Lath. suppl.* 2.

Size of a turtle: plumage wholly green, except the tips of the quills, which are brown, and the orbits fulvous: tail very long and cuneiform. Native of Chili, where it is often seen in large flocks.

BLACK-NECKED PARRAKEET.

Psittacus nigricollis. *P. macr. viridis, collo subtus nigro, loris lineaque lateribus colli albis. Lath. ind. orn. suppl.* 2.
Green Parrakeet, with the fore-part of the neck black; the lores and line on each side the neck white.
Black-Necked Parrakeet. *Lath. syn. suppl.* 2.

Size of Alexandrine Parrakeet: native of Brasil: described by Dr. Latham from a drawing in the collection of General Davies.

CRIMSON-VENTED PARRAKEET.

Psittacus erythropygius. *P. macr. viridis, capite colloque flavis, crisso coccineo, remigibus rectricumque apice cæruleis. Lath. ind. orn.*
Green Parrakeet, with yellow head and neck, scarlet vent, blue quill-feathers, and tail tipped with blue.
Crimson-vented Parrot. *Lath. syn.*

Size rather large: bill dusky: tail cuneiform: vent crimson: quills and end of the tail-feathers blue. Described by Dr. Latham from a specimen in the Leverian Museum. Supposed to be a native of the East Indies.

CAROLINE PARROT.

Psittacus Carolinensis. *P. macr. viridis, capite colloque luteis.*
Green Parrakeet, with orange-yellow head and neck.
Psittacus Carolinensis. *Lin. Syst. Nat.*
Caroline Parrot. *Lath. syn.* *Catesb. Car.* 1. *pl.* 11.

THE size of this species is similar to that of the Sincialo Parrakeet, and its length about thirteen inches: forehead, side of the wings, and feathers round the knees orange: head and neck yellow: larger quill-feathers dusky, mixed with blue and green; the upper exterior sides edged with yellow: tail very long; bill and legs white. Native of Guiana and other parts of South America, migrating in autumn into various parts of North America, particularly Carolina and Virginia; arriving at the season when mulberries are ripe, which it is very fond of. It also commits great havoc among the apples, splitting the fruit, according to Catesby, merely for the sake of the kernels. It also feeds on the seeds of the Cypress and other trees, as well as on the buds of the birch, &c. Very few of these birds remain in Carolina during the whole year. They breed in hollow trees, in low, swampy grounds, and when taken are easily tamed, but cannot be taught to speak.

As a proof how common this species is in Carolina and Virginia, Monsr. Levaillant assures us that he has seen a packet of above six thousand skins of this bird sent to a Plumassier at Paris for the purpose of ornamenting dresses, &c.

ILLINOIS PARRAKEET.

Psittacus pertinax. *P. macr. viridis, genis fulvis, remigibus rectricibusque canescentibus. Lin. Syst. Nat.*

Green Parrakeet, with fulvous cheeks, and greyish quill and tail-feathers.

Illinois Parrot. *Lath. syn.*

Yellow-faced Parrakeet. *Edw. pl.* 234.

SIZE similar to that of the Caroline Parrakeet; from which it has been sometimes considered as differing merely in sex: forehead, cheeks, and whole head bright orange-colour, but the crown generally green: upper parts of the body, wings, and tail green: larger quill-feathers green, edged with blue: breast and belly yellowish-green, vent yellow: bill pale; legs deep ash-colour. Native of the Brasils, and other parts of South America, as well as of the interior parts of North America, as in the country of the Illinois, south of the Lake Michigam, and in its manners resembling the Caroline Parrakeet.

Mr. Pennant, who describes this bird in his Arctic Zoology, observes that the Count de Buffon confines the whole Parrot genus to exactly twenty-five degrees on each side the Equator; but that this limit is unquestionably by far too contracted. In his Index to the Planchés Enluminées, Mr. Pennant is somewhat more pointed in his observations, and I shall extract what he has said on the subject. Indeed, after the numerous species of Parrots which have been discovered in New Holland, we can no longer doubt the impropriety of the limits assigned to this genus by Buffon.

"This illustrious author, says Mr. Pennant, having resolved that no Parrots should pass beyond the tropic of Capricorn, despises the authority of the Dutch navigator Spilbergen, who was eye-witness to the woods of Terra del Fuego, the very southern boundary of the Straits of Magellan, in lat. 44, being full of a species of these birds. He might have cited the evidence of Captain Hood, who saw a small Parrot at Cape Famine; and he might have quoted Commodore Byron, who says that, notwithstanding the coldness of the climate, he observed Parrots innumerable in the woods of the same harbour. Mr. Edwards, one of the surgeons, now living at Caernarvon, informed me that he saw them in abundance, and that they were of a deep green; probably the very species engraven in the Planches Enluminées, No. 85. The Count treats with the same contempt the authority of the observant and veracious Captain Cook, who, in defiance of the Count's canon, had the hardiness to trust to the evidence of his own senses, and assert that he saw Parrots in the isle of New Zealand; and even to suffer Captain Furneaux to blab out that Parrakeets were inhabitants of Van Diemen's Land, the very extremity of New Holland; both of them countries interdicted by the illustrious naturalist to the whole Parrot race. How greatly again has our able navigator aggravated matters by not silencing the learned Forster, for proving more than one species to be found in the raw wet climatė of Dusky Bay, in lat. 46; and to make bad worse, to connive at several of the companions of his voyage bringing into this kingdom not fewer than eight species of this vagabond genus, who had dared to take up their residence beyond the genial limits of the torrid zone, which the Count de Buffon had so authoritatively decreed to them, and, like a great Creator, had said, hitherto shalt thou come, and no farther!"

The Illinois Parrakeet varies considerably in plumage from age and other circumstances. The male, according to Levaillant, has a yellow face, and a rufous-grey breast; blue quill-feathers, and the tail about or nearly the length of the body. The female is rather smaller than the male, and has a shorter tail in proportion. The young of both sexes resemble each other completely, and have no yellow on the face, which is rufous-grey.

Mr. Levaillant has observed many varieties, which, having been kept in a state of domesticity, were more or less covered with yellow on different parts of the plumage: one in particular which had the whole under parts, from the bill to the tail, of a fine deep yellow; and another which had the front and breast of that colour.

SOLSTITIAL PARRAKEET.

Psittacus solstitialis. *P. macr. luteus, alarum tectricibus viridibus, orbitis rubris, rectricibus lateralibus extus cæruleis. Lin. Syst. Nat.*

Luteous Parrakeet, with green wing-coverts, red orbits, and tail-feathers edged outwardly with blue.

Angola Yellow Parrot, and Brasilian Yellow Parrot. *Lath. syn.*

Psittacus aurantius. *Miller Ill. pl. 5.*

Perruche jaune, et Perruche Guarouba. *Buff. ois.*

Perruche jaune de Cayenne. *Pl. Enl. 525.*

SOMEWHAT smaller than the common Parrakeet: length about thirteen inches: colour fine yellow, more or less clouded in different specimens with bright orange colour: wing-coverts slightly varied with green: larger quill-feathers blue, some of them edged and tipped with yellow-green: tail apple-green, with blue tips: orbits of the eyes bare and flesh-coloured; bill dusky; legs flesh-colour.

There can be no doubt that the bird figured in the Planches Enluminées, No. 525, under the title of Perruche jaune de Cayenne, represents the present species, but in a young state, as Monsr. Levaillant observes. If collated with the figure in Miller's miscellaneous plates of Natural History, it will be found to differ merely in being of a somewhat paler yellow, or with less of the orange cast, and with the tail somewhat shorter in proportion. Native of Brasil, and not, as formerly supposed, of Angola in Africa.

The female is rather smaller, according to Monsr.

SOLSTITIAL PARRAKEET.

1811, Aug.t 1. London, Published by G. Kearsley, Fleet Street, & the other Proprietors.

Levaillant, and is of a jonquil yellow on the upper parts, with orange front, cheeks, and sides of the body; and the wings have more green and less blue in them than those of the male: the same is also the case with the tail, which is only slightly edged and tipped with blue. The young or first year's male is of a still paler yellow than the female, with green wing-feathers, slightly tinged with blue, and the tail the same. It is the state in which this species is represented in the Planches Enluminées, No. 525.

The Solstitial Parrakeet is twice described by Buffon, under different divisions of the genus; viz. first among the cuneated *equal-tailed* division, under the name of Perruche jaune; and again under the *unequal-tailed*, under the name of Perruche Ara Guarouba.

VAR. ?

BRASILIAN YELLOW PARRAKEET.

Smaller than the common Parrakeet, measuring about eleven inches: colour of the whole bird fine yellow, except the greater quill-feathers, which are green: bill and legs grey. Native of Brasil and some other parts of South America. It is said to be of a solitary nature, and is easily tamed, but does not learn to articulate, being chiefly valued on account of its colour and scarcity. The account given by the Count de Buffon of his Guarouba ou Perruche jaune can apply to no other than the preceding species, which the plate referred to in the Planches Enluminées evidently represents.

VAR.

MEXICAN YELLOW PARRAKEET.

In size it is equal to the common Parrakeet, and measures about eleven inches in length: the head is pale red, the neck orange-coloured: the back, rump, breast, belly, sides, thighs, and under tail-coverts are light yellow: the upper wing-coverts varied with green, red, and orange: the quill-feathers green, the tail pale yellow, and the bill reddish. Native of Mexico.

I conclude the account of the Solstitial Parrakeet by proposing a query, viz. whether the Illinois and Caroline Parrakeets may not in reality be varieties of the same species, viz. the Solstitial Parrakeet, differing only in sex, and the more or less advanced state of their respective plumage?

JENDAYA PARRAKEET.

Psittacus Jendaya. *P. macr. viridis, capite collo abdomineque flavis. Lath. ind. orn.*
Green Parrakeet, with yellow head, neck, and belly.
Jendaya. *Will. orn. Buff. ois.*
Yellow-Headed Parrot. *Lath. syn.*

Slightly described by Marcgrave, who says it is of the size of a Blackbird, with the head and neck yellow, and the plumage green above, and yellow beneath, with the bottom of the belly green, and the bill and legs black: the shape of the tail is not particularized, so that even the section to which it belongs in the genus remains uncertain. Native of Brasil, where it is called Jendaya. Perhaps no other than a variety of the preceding species.

BLACK PARRAKEET.

Psittacus Novæ Guineæ. *P. macr. cæruleo-nigricans nitidus, orbitis nudis fuscis, cauda subtus rubra. Lath. ind. orn.*
Glossy blue-black Parrakeet, with naked brown orbits, and tail red beneath.
Black Lory. *Lath. syn.* Lori noir, &c. *Sonner. voy. Levaill. pl.* 49.

Described by Sonnerat: size of the common Parrakeet: colour black, with a cast of blue, accompanied by a metallic gloss, and a velvet softness: tail red beneath, growing yellowish towards the tips: bill and legs blackish. Native of New Guinea.

HORNED PARRAKEET.

Psittacus bisetis. *P. viridi-flavescens, capite (maris) coccineo pennis duabus elongatis, alis caudaque extus cærulcis.*

Yellowish-green Parrakeet, with the head (of the male) scarlet and furnished with two lengthened feathers, and the wing and tail-feathers outwardly blue.

Psittacus bisetis. Horned Parrot. *Lath. male.*

Psittacus Caledonicus. Caledonian Parrot. *Lath. female.*

Length about eleven inches: habit that of the Alexandrine Parrakeet. Bill and legs blueish-black. In the female the frontlet alone is crimson. Native of New Caledonia.

CRESTED PARRAKEET.

Psittacus Novæ Hollandiæ. *P. fusco-olivaceus, capite (maris) luteo pennis senis cristato, macula postoculari coccinea, fascia alari alba.*

Brownish-Olive Parrakeet, with the head, (in the male) luteous and furnished with a six-feathered crest, a crimson spot behind each eye, and a white bar on the wings.

Psittacus Novæ Hollandiæ. Crested Parrakeet. *Lath.*

Length about twelve inches: in the female the head is olive, but crested as in the male; and the rump and tail crossed by numerous grey bars. Native of New Holland.

BONTIAN PARRAKEET.

Psittacus Bontii. *P. macr. coccineus, gula grisea, alis et scapularibus viridi variis, rectricibus lateralibus roseis apice cœruleis. Lath. ind. orn.*

Scarlet Parrakeet, with grey throat, wings and scapulars varied with green, and the outside tail-feathers rose-coloured with blue tips.

Psittacus parvus Bontii. *Ray. et Will.* Crested red Parrakeet, *Lath.*

Described by Bontius in his History of Java: size of a Lark: tail very long: bill and legs grey. It builds in lofty trees, flies in flocks; in confinement is very garrulous, and may be taught to articulate.

GOLDEN-CROWNED PARRAKEET.

Psittacus Regulus. *P. viridis, vertice aurantio, fascia alarum remigibusque extus cœruleis.*

Green Parrakeet, with orange crown, blue bar across the wings, and tail-feathers blue on the outsides.

Golden-Crowned Parrakeet. *Edw. pl.* 235.

Psittacus Brasiliensis. *Lath. ind. orn.*

Size, according to Edwards, its first describer, rather larger than a Blackbird: throat yellowish, with a tinge of dull red: bill black: orbits bare and flesh-coloured. Native of Brasil.

GROUND PARRAKEET.

Psittacus terrestris. *P. viridis nigro flavoque maculatus, capistro rubro, cauda flava fasciis numerosis nigris. Zool. New Holl. No.* 1. *pl.* 1. *Nat. Misc. vol.* 7. *pl.* 228. *Museum Leverianum, p.* 217.

Green Parrakeet, spotted with black and yellow; with red frontlet, and yellow tail crossed by numerous black bars.

Psittacus formosus. *Lath: ind. orn.*

La Perruche ingambe. *Levaill. pl.* 32.

A HIGHLY elegant and singular species: size that of a Turtle; ground-colour green, brightening almost into yellow on the back, and yellow-green beneath: whole plumage on the upper parts beautifully variegated with black, each feather being crossed by three or four bars of that colour: larger wing-feathers plain green, inclining to blue on the edges; some of the interior ones marked by a yellow spot, and forming a yellow bar when the wings are spread: forehead as far as the eyes red: tail strongly cuneated, and consisting of sharp-pointed feathers of the finest jonquil yellow, crossed by very numerous arrow-shaped black bars, the forks pointing downwards; the two middle feathers green, with numerous paler bars of the same shape: bill and legs horn-coloured; the structure of the feet remarkable, the toes being of a more slender and lengthened shape than in the rest of the Parrot tribe, and the exterior hind-claw remarkably long. In its habits also this bird departs from the rest of

GROUND PARRAKEET.

1811, Aug. 5. London, Published by G. Kearsley, Fleet Street, & the other Proprietors.

the genus, being hardly ever observed to perch on trees, but to remain on the ground in sedgy places, or among long grass, and running in the manner of a Rail. Native of New Holland.

RED-AND-BLUE-HEADED PARRAKEET.

Psittacus canicularis. *P. viridis, subtus subflavescens, fronte rubra, vertice cæruleo, orbitis nudis aurantiis.*

Green Parrakeet, yellowish-green beneath, with red forehead, blue crown, and naked orange-coloured orbits.

Psittacus canicularis. *Lin. Lath.*

Red-and-Blue-Headed Parrakeet. *Edw. pl.* 176.

Buff-fronted Parrakeet. *Lath. syn.*

First described by Edwards: size of a Thrush: length about ten inches: larger quill-feathers tipped with blue: tail strongly cuneated: bill pale grey; legs dull flesh-colour. Native of South America.

BROWN-THROATED PARRAKEET.

Psittacus æruginosus. *P. viridis, vertice remigibusque primoribus cæruleis, genis gulaque fusco-cinereis.*

Green Parrakeet, with blue crown and larger quill-feathers, and grey-brown cheeks and throat.

Psittacus æruginosus. *Lin. Lath.*

Brown-Throated Parrakeet. *Edw. pl.* 177.

Length about ten inches: larger quill-feathers tipped with blue: bill grey: legs brownish flesh-colour: tail of moderate length. Native of South-America: sometimes varies, in having the fore-head brown.

GREY-BREASTED PARRAKEET.

Psittacus murinus. *P. olivaceus remigibus viridioribus, facie gula pectoreque griseis.*

Olive Parrakeet, with greener quill-feathers, and grey face, throat, and breast.

Psittacus murinus. Grey-Breasted Parrakeet. *Lath.*

La Perruche souris. *Buff. ois.*

LENGTH ten inches: bill and legs grey. Native of South-America, and said to be of a docile nature, easily tamed, and taught to speak.

LITTLE RED-WINGED PARRAKEET.

Psittacus incarnatus. *P. viridis, gula tectricibusque alarum rubris, rostro pedibusque incarnatis.*

Green Parrakeet, with red throat and wing-coverts, and flesh-coloured bill and legs.

Little Red-Winged Parrakeet. *Edw. pl.* 236.

Psittacus incarnatus. *Lath. ind. orn.*

SIZE of a Crested Lark: length about eight inches and a half: chin scarlet; wing-coverts paler red. Native of the East Indies.

SOCIETY PARRAKEET.

Psittacus Ulietanus. *P. macr. capite fusco, uropygio obscure coccineo, corpore viridi-olivaceo, marginibus pennarum remigibus rectricibusque nigricantibus. Lath. ind. orn.*

Dusky-Olive Parrakeet, with brown head, dull-red rump, dusky quill and tail-feathers, and those of the remainder of the plumage edged with dusky.

Society Parrot. *Lath. syn.*

Length ten inches and a half: bill blue; legs black. Native of the Society Isles.

BIMACULATED PARRAKEET.

Psittacus bimaculatus. *P. macr. viridis, maculis colli duabus oblongis nigris, utriusque alæ macula magna et sulphurea. Lath. ind. orn. Sparm. Mus. Carls.*

Green Parrakeet, with two oblong black spots on the neck, and a large sulphur-coloured spot on each wing.

Length about eleven inches: front and throat inclining to orange-colour: on each side the neck a large black spot composed of several small ones: edges of the quill-feathers pale yellow, forming a large spot on the wing. Native region unknown.

YELLOW-WINGED PARRAKEET.

Psittacus virescens. *P. virescens, remigibus primoribus cærulescentibus, secundariis albis flavo marginatis, tectricibus flavo terminatis.*

Palish-green Parrakeet, with the greater quill-feathers blueish, the secondaries white with yellow edges, and their coverts tipped with yellow.

Psittacus virescens. Yellow-Winged Parrakeet. *Lath.*

La Perriche à ailes variées. *Buff. ois. Pl. Enl.* 359. *Levaill. pl.* 57.

Size less than that of a Blackbird: length about eight inches and a quarter: wings varied longitudinally with pale yellow, white, and blueish-green: bill and legs grey: tail cuneated, but rather short than long in proportion to the bird. Native of Cayenne, where, according to Buffon, it is not uncommon, often flying in flocks, and even settling in the midst of inhabited places. Monsr. Levaillant seems to confound this species with the Gold-Winged Parrakeet of Edwards, or Psittacus chrysopterus of Linnæus, belonging to the second division of the genus.

LINEATED PARRAKEET.

Psittacus lineatus. *P. macr. viridis, remigibus subtus fuscis margine interiore pallidis. Lin. Syst. Nat.* 3. *append. p.* 223.
Green Parrakeet, with the quill-feathers brown beneath, with pale inner margins.
Lineated Parrot. *Lath. syn.*

Size of a dove; the wings brown beneath, with the inner margins of the feathers very pale, causing a longitudinally striped or lineated appearance: tail rather longer than body. Native region uncertain.

SEBAN LORY.

Psittacus Sebanus. *P. cæruleus subtus ruber, pileo nigro, collari rubro viridique, abdomine femoribusque viridibus.*
Blue Lory, red beneath, with black crown, red and green collar, and green belly and thighs.
Psitt. orient. exquis, Loeri dictus. *Seb.* I. *p.* 63. *t.* 38. *f.* 4.
Psittacus atricapillus. Black-crowned Parrot. *Lath.*

Slightly described, and figured in Seba's Thesaurus. Length seemingly about fourteen inches: size rather larger than that of the Amazons Parrot: bill pale, or yellowish-white; legs black. The individual mentioned by Seba was a native of one of the Papuan isles: it had been taught to speak in various languages with great distinctness, and was of a mild and affectionate disposition.

BORNEAN LORY.

Psittacus Borneus. *P. macrourus ruber, remigibus rectricibusque apice viridibus, alis macula cærulea, orbitis fuscis. Lin. Syst. Nat.*

Red Lory, with the quill and tail-feathers tipped with green; a blue spot on the wings, and dusky orbits.

Long-tailed Scarlet Lory. *Edw. pl.* 173. *Lath.*

La Perruche ecarlate. *Levaill. pl.* 44.

It has been usual among collectors of birds to call such of the middle-sized Parrots and Parrakeets as have a red ground-colour by the name of Lories. It must be observed however that this is by no means a scientific, but merely a popular distinction. It should also be added, that species of this denomination rank in either division of the genus according to the different proportion and shape of the tail, which in some is lengthened and cuneated, and in others short and even. The present beautiful species is from nine or ten to twelve inches in length, and is of a bright, but rather light scarlet colour, deeper or purplish on the back and tail, and somewhat inclining to yellow on the breast: the tips of the larger wing-coverts, as well as those of the shorter, and sometimes of the longer quill-feathers, together with those of the tail, are green: the scapulars fine pale blue: the tail is moderately cuneated, the orbits of the eyes bare and dusky; the bill pale orange, and the legs brown. Native of the island of Borneo, where it is not uncommon.

M. Griffith sculp.

BORNEAN LORY.

1811 Aug. London, Published by G. Kearsley, Fleet Street, & the other Proprietors.

INDIAN LORY.

1811, Aug.t 1 London, Published by G. Kearsley, Fleet Street, & the other Proprietors.

INDIAN LORY.

Psittacus coccineus. *P. coccineus violaceo undulatus, vertice collo superiore abdomine caudaque cæruleis, remigibus fusco-flavescentibus.*

Crimson Lory, with violet undulations, blue crown, upper-neck, abdomen, and tail, and yellowish-brown quill-feathers.

Psittacus coccineus. Indian Lory. *Lath.*

Lori Perruche violet et rouge. *Buff. ois. Levaill. pl.* 55.

Perruche des Indes orientales. *Pl. Enl.* 143.

A MOST elegant species: length near eleven inches; colour fine deep scarlet or crimson, variegated with violet-blue, as described in the specific character: larger wing-feathers yellowish-brown: bill red: legs blackish. Native of the Molucca islands. It may not be improper to observe that in the Planches Enluminées of Buffon the quill-feathers are coloured yellow.

BLUE-HOODED LORY.

Psittacus cucullatus. *P. coccineus, nucha collo abdomineque cæruleis, remigibus violaceo-olivaceis rubro mistis.*

Crimson Lory, with blue nape, neck, and belly, and violet-olive quill-feathers mixed with red.

Perruche à chaperon bleu. *Levaill. pl.* 54.

THIS species Monsr. Levaillant is inclined to think entirely new or before undescribed. Size rather smaller than that of the Indian Lory:

colour deep plain scarlet or crimson, the back-part of the head and neck deep or violet-blue, passing forwards and entirely surrounding the neck with a broad collar: on the middle of the belly a very large patch of the same colour: greater and smaller quill-feathers deep violet-olive, or nearly blackish, marked by an interrupted red bar: tail of moderate length, and slightly cuneated: bill red: legs black. Native of the Moluccas.

GUEBY LORY.

Psittacus Gueoiensis. *P. macr. coccineus, pectore, interscapuliis abdomineque purpureis, hoc saturatiore, remigibus nigris fascia rubra. Lath. ind. orn.*

Scarlet Lory, with purple breast, interscapulars, and belly, the latter darkest, and black quill-feathers with a red bar.

Perroquet violet. *Fermin Surin. 2. p. 175.*

Dr. Latham's description of this species seems to come very near to the immediately preceding bird, more especially as he mentions a broad ring of purple, beginning at the lower part of the neck behind, and passing to the breast, where it occupies a large space. Some of the synonyms however quoted by Dr. Latham seem by no means to agree with his specific character, but rather with the following bird, which yet may be the same species in a different stage of colour.

SCALY LORY.

Psittacus squameus. *P. coccineus, nigricante undulatus, alis subnigris coccinea variis, cauda fusco-rubente.*
Crimson Lory, with blackish undulations, blackish wings varied with crimson, and brownish-red tail.
Loris rouge et violet. *Buff. ois.*
Loris de Gueby. *Pl. Enl.* 684. *et var.?* 519.
Le Lori écaillé? *Levaill. pl.* 51.
P. ruber? & Moluccensis? *Lath. ind. orn.*
Molucca Lory? *Lath. syn.*

LENGTH nine or ten inches: colour bright crimson or carmine; the plumage of the back of the head, neck, breast, and belly beautifully undulated or scalloped with violet-colour: wings and scapulars blackish-violet, the wings barred and scalloped with crimson: tail dull or brownish-crimson, of moderate length, and moderately cuneated: bill red: legs dusky. Native of Gueby and the rest of the Moluccas. The Lori écaillé of Monsr. Levaillant seems to be the same species, differing merely in cast of colour, the crimson being rather deeper, and the wings rather dusky-olive than violet, but varied in a similar manner. The *Lori de la Chine* of the Pl. Enl. No. 519. differs in having the plumage plain crimson, without undulations, and the tips alone of the larger and smaller quill-feathers violet or blackish.

CRIMSON LORY.

Psittacus puniceus. *P. coccineus, subtus violaceus, rectricibus saturate coccineis apice rubescentibus. Lath. ind. orn.*

Crimson Lory, violet beneath, with deep crimson tail pale red at the tip.

Le Loris cramoisi. *Buff. ois.* Lori d'Amboine. *Pl. Enl.* 518.

Blue-breasted Parrot. *Brown. Illustr. pl.* 6.

LENGTH eleven inches and a half: colour deep crimson, brightest on the head and breast: edge of the shoulders violet: greater quill-feathers violet-blue: tail long, yet but very slightly cuneated: bill reddish; legs brown. Native of the Moluccas.

VARIEGATED LORY.

Psittacus variegatus. *P. macr. coccineus, dorso anteriore corporeque subtus purpureo-cæruleis, remigibus intus flavis, cauda viridi. Lath. ind. orn.*

Crimson Lory, with the fore-part of the back, and the body beneath violet-blue, the quill-feathers yellow within, and the tail green.

THIS bird was preserved in the Leverian Museum, and was rather smaller than the Bornean or Scarlet Lory; its length was near eleven inches, and its general colour deep scarlet, but the lower or back part of the neck, upper part of the back, breast, belly, and vent fine purplish blue, but deepening almost to blackness on the lower part of the belly; the tail green, but slightly cuneated, and the two exterior feathers edged towards the tips with blue: bill dusky. Native of the East Indies.

ZONED PARRAKEET.

Psittacus viridis. *P. capite remigibusque nigris, torque nuchali fasciaque abdominali flavis.*

Green Parrakeet, with black head and quill-feathers, and yellow nape-collar and abdominal zone.

Psittacus zonarius. Zoned Parrot. *Nat. Misc.* 16. *pl.* 657.

DESCRIBED in the Naturalist's Miscellany: size that of the Tabuan Parrakeet: tail long, and strongly cuneated, but not acute at the end: bill brown; legs dusky. Native of New Holland.

DUBIOUS PARRAKEET.

Psittacus dubius. *P. macr. viridis, collo rufescente, remigibus toto rectricibusque quatuor intermediis apice cæruleis, orbitis nudis flavescentibus. Lath. ind. orn.*

Green Parrakeet, with blue-edged quill-feathers, the four middle tail-feathers tipped with blue, and yellowish naked orbits.

LENGTH nine inches: bill and legs dusky: edge of the alulet blue. Native region uncertain.

RED-HOODED PARRAKEET.

Psittacus capitatus. *P. olivaceo-flavescens, capite pectoreque rubris, remigibus caudaque cæruleis.*

Yellow-olive Parrakeet, with red head and breast, and blue quill and tail-feathers.

Lori Perruche de la mer de Sud. *Sonnini Buff.*

Described by Mons. Virey in Sonnini's edition of Buffon. Length between seven and eight inches: bill and legs grey. Native of the Southern islands.

RED-SHOULDERED PARRAKEET.

Psittacus discolor. *P. viridis, humeris margine subtusque sanguineis, tectricibus cæruleis, rectricibus basi ferrugineis.*

Green Parrakeet, with the margins and under parts of the shoulders red, the wing-coverts blue, and the tail-feathers ferruginous at the base.

Red-shouldered Paroquet. *White's Journ. p. 263.*

La Perruche Latham. *Levaill. pl. 62. & Perr. Banks. pl. 50?*

Described and figured in White's Journal of a Voyage to New South Wales: length about ten inches: bill and legs pale brown. Native of New Holland.

BLACK-BACKED PARRAKEET.

Psittacus melanotus. *P. viridis, dorso nigro, tectricibus alarum sanguineis, uropygio cæruleo.*

Green Parrakeet, with black back, scarlet wing-coverts, and blue rump.

Psittacus melanotus. *Nat. Misc.* 16. *pl.* 653.

Psittacus erythropterus? *Lath. ind. orn.*

A most beautiful species: length eleven or twelve inches: size of a small turtle: colour very bright yellowish grass-green, somewhat paler beneath: all the wing-coverts of the brightest scarlet: scapulars jet-black: smaller quill-feathers black, but appearing in a full light very dark green: tail rather long, but so slightly cuneated as to leave it doubtful whether it should not rather be placed in the even-tailed division of the genus, and slightly tipped with pale yellow: bill red: legs dusky. Native of New Holland.

ORANGE-BELLIED PARRAKEET.

Psittacus chrysogaster. *P. macr. viridis, tectricibus alarum majoribus extus cæruleis intus macula alba, abdomine postico aurantio, rectricibus quatuor extimis apice luteis. Lath. ind. orn.*

Green Parrakeet, with the larger wing-coverts blue, marked within by a white spot, lower belly orange-coloured, and the four outward tail-feathers tipped with yellow.

Described by Dr. Latham from Mr. Pennant. Length seven inches and a half: bill and legs greenish. Native country uncertain.

PALM PARRAKEET.

Psittacus palmarum. *P. submacr. viridis, rostro pedibusque rubentibus, rectricibus tribus exterioribus utrinque margine apiceque flavis. Lath. ind. orn.*

Green Parrakeet, with reddish bill and legs, and the three exterior tail-feathers edged and tipped with yellow.

Length about eight inches: tail sharp, but not very long. Native of the Southern isles, and particularly that of Tanna, frequenting palm-trees.

PEREGRINE PARRAKEET.

Psittacus peregrinus. *P. macr. viridis, macula alarum longitudinali fusca. Lath. ind. orn.*

Green Parrakeet, with a longitudinal brown band.

Length eight inches: tail short, but cuneiform: bill and legs pale red. Native of the Southern isles.

UNDULATED PARRAKEET.

Psittacus undulatus. *P. viridi-flavescens, supra fusco undulatus, cauda cærulea fascia utrinque versus basin obliqua flavissima.*

Yellowish-green Parrakeet, undulated above with brown, with blue tail crossed on each side towards the base by an oblique deep-yellow bar.

Psittacus undulatus. *Nat. Misc.* 16. *pl.* 673.

Described in the Naturalist's Miscellany: length about seven inches and a half: cheeks marked by a few small black spots: tail strongly cuneated: bill and legs pale brown. Native of New Holland.

Var?

Of a palish sparrow-colour above, with greenish quill-feathers, tail, and under parts: head very finely traversed with numerous tender dark lines or streaks, and on, or rather under each cheek, a small ultramarine-blue spot, formed by the tips of three or four of the feathers. In the Liverpool Museum; received from New Holland.

TURCOSINE PARRAKEET.

Psittacus pulchellus. *P. viridis, subtus luteus, capistro alisque cœruleis.*

Green Parrakeet, yellow beneath, with blue wings and frontlet. *Nat. Misc.* 3. *pl.* 96.

La Perruche Edwards. *Levaill. pl.* 68.

First described in the Naturalist's Miscellany: length from six to seven inches: side-feathers of the tail yellow, with a black streak near the shaft; the two middle feathers green: bill blackish; legs brown. Native of New Holland, and a most elegant species, representing, in some degree, the Psittacus Ararauna in miniature.

ORANGE-WINGED PARRAKEET.

Psittacus pyrrhopterus. *P. submacr. viridis, vertice cœrulescente, humeris tectricibusque alarum inferioribus aurantiis. Lath. ind. orn. suppl.*

Green Parrakeet, with blueish crown, and orange shoulders and under-coverts.

Length seven or eight inches: bill pale; legs red: supposed to inhabit the Brasils.

SMALL PARRAKEET.

Psittacus pusillus. *P. viridis, capistro rectricumque basi rubris.*
Green Parrakeet, with the frontlet and base of the tail-feathers red.
Small Parrakeet. *White's Journ. p. 262.*

LENGTH between six and seven inches: tail strongly cuneated, with the side-feathers red on their inner webs towards the base: bill brownish; legs dusky: tongue furnished at the tip with a kind of brush of lengthened papillæ, as in several of the smallish New Holland Parrakeets.

PALE PARRAKEET.

Psittacus pallidus. *P. flavus, alis albidis rosaceo leviter tinctis, virescenteque, pro luce, variatis. Nat. Misc. pl.* 258.
Yellow Parrakeet, with whitish wings slightly tinctured with rose-colour, and with a cast of greenish according to the direction of the light.

A SMALL species, measuring about six inches in length: tail slightly cuneated, but rather short: bill and legs whitish: supposed to be a native of New Holland.

BLUE-CRESTED PARRAKEET.

Psittacus pipilans. *P. viridis, vertice subcristato cæruleo, gula abdomineque medio rubris.*

Green Parrakeet, with slightly-crested blue crown, red throat, and red abdominal bar.

Psittacus pipilans. *Lath. ind. orn.* Psittacus porphyrocephalus. *Nat. Misc. pl.* 1.

La Perruche fringillaire. *Levaill. pl.* 71.

A most beautiful species: length about six inches: feathers of the crown narrow and pointed: tail short, but pretty strongly cuneated: bill and legs orange-red. Native of the Sandwich islands.

PHIGY PARRAKEET.

Psittacus coccineus. *P. vertice femoribusque violaceis, alis cauda lunulaque cervicali viridibus.*

Crimson Parrakeet, with violet crown and thighs, and green wings, tail, and nuchal crescent.

La Perruche Phigi. *Levaill. pl.* 64.

Allied in appearance to the Blue-crested Parrakeet, but seems a degree larger: bill reddish; legs yellowish: described by Levaillant from a specimen in the Paris Museum.

BLUE CRESTED PARRAKEET.

1811. Aug.t 1 London, Published by G. Kearsley, Fleet Street.

VIOLET PARRAKEET.

1811. Aug. 1. London, Published by G. Kearsley, Fleet Street, & the other Proprietors.

VIOLET PARRAKEET.

Psittacus Porphyrio. *P. violaceo-cæruleus, vertice subcristato, gula alba.*
Violet-blue Parrakeet, with slightly-crested crown, and white throat.
Psittacus Taitanus. *Lath. ind. orn.*
Psittacus Porphyrio. Violet-Blue Parrakeet. *Nat. Misc. pl. 7.*
La Perruche Arimanon. *Levaill. pl. 65.*
La Perruche Sparman. *Levaill. pl. 66.*

A size smaller than the Blue-crested Parrakeet, which it resembles in shape: sometimes varies in having no white on the throat: bill orange; legs flesh-colour. Native of some of the small Southern islands, and particularly of Otaheitee, feeding on the fruit of the Banana or Musa; the structure of the tongue resembling that of several other Southern Parrakeets, being furnished with a sort of fibrous brush at the tip.

PYGMY PARRAKEET.

Psittacus pygmæus. *P. submacr. viridis, pennarum apice ex virescente flavo, remigibus intus obscuris. Lath. ind. orn.*
Green Parrakeet, with the feathers tipped with yellow-green, and the quill-feathers dusky within.
Pygmy Parrakeet. *Lath. syn.*

Length six inches: bill whitish; legs lead-colour. Native of the Sandwich islands: first described by Dr. Latham, from a specimen in the Leverian Museum.

With more or less short and even tails.

BLACK COCKATOO.

Psittacus Gigas. *P. cristatus nigro-cinereus, (femina dilutiore,) genis nudis rubris.*

Crested Grey-black Parrot, (the female paler,) with naked red cheeks.

Psittacus Gigas. *P. brach: niger, crista occipitis dilutiore elongata, genis nudis rubris. Lath. ind. orn.*

Le Kakatoës noir. *Buff. ois.*

Great Black Cockatoo. *Edw. pl.* 316.

L'Ara noir a trompe.
L'Ara gris a trompe. } *Levaill. pl.* 11, 12, 13.

THE Black or Giant Cockatoo was first described and figured by Edwards, from a drawing taken from the living bird in the island of Ceylon, and communicated by Governor Loten. Edwards's description runs as follows. " This is a parrot of the first magnitude: it equals in size the Great Scarlet, and the Blue and Yellow Maccaw: the bill is strong, much arched or hooked, and of a dusky brown colour, with an angle on each side of the upper mandible: the nostrils appear to be amongst the feathers at the basis of the upper mandible: the eyes are of a dark colour: the sides of the head, from the eyes to the under part of the bill, are covered with a red, bare, wrinkled

BLACK COCKATOO.

1811 Aug.t 1. London Published by G. Kearsley Fleet Street, & the other Proprietors.

skin: the crest is of a light grey colour, the feathers pointed and reflected upwards at their tips: the bird can raise it higher, or let it fall flat on the head: the whole plumage, from the crest downwards, is of a blueish black or dark lead-colour, a shade lighter on the under side than on the back and wings: the tail is longer than usual, where the feathers are of equal length: the legs and feet are of a brownish black, like the bill, covered with a rough scaly skin."

Edwards observes that this species seems to be figured in a small book of prints of birds, drawn from the life, and published by S. Vander Meulen at Amsterdam, 1707. It is there called by the name of *Corvus Indicus.*

The female seems to be described in Mons. Levaillant's splendid History of Parrots, under the title of *Aras gris a trompe.* It resembles the male, except in its colour, which is of a slate-grey. Monsr. Levaillant observes that the tongue of this species of Cockatoo has a great peculiarity in its structure; being hollow at the tip, so as to constitute a kind of tube or trunk, assisting the bird in the act of taking its food, and of penetrating into the substance of fruits, &c. Mons. Levaillant also observes that this bird, in cold weather, covers the bare space on each side the bill, by lowering over them the plumes of the crest.

BANKSIAN COCKATOO.

Psittacus Banksii. *P. subscristatus niger, fronte humerisque gilvo maculatis, cauda medio rubra fasciis numerosis nigris.*

Slightly-crested black Cockatoo, with buff-spotted front and shoulders, and tail red in the middle with numerous black bars.

Banksian Cockatoo. *Lath. Syn. suppl. pl.* 109. *White's Journ. pl. p.* 139.

Magnificent Cockatoo. *Nat. Misc.* 2. *pl.* 50.

Funereal Cockatoo. *Nat. Misc.* 6. *pl.* 186.

The Banksian Cockatoo is a native of New Holland, where it was discovered during the first circumnavigation of Captain Cook. Its size is that of the Scarlet Maccaw, and its length about twenty-two inches: its colour is black, more or less varied or spotted about the shoulders, front, and under parts of the body with yellowish or clay-coloured spots: the head is furnished with a moderately large crest, but not extending so far back as in most of the other cockatoos; and the tail, which is of greater length than in the rest, is bright scarlet; all the feathers except the two middle ones being crossed by pretty numerous black bars, making an elegant appearance when the tail is spread: the bill is very large, and of a yellowish horn-colour, and the legs black.

This magnificent species varies considerably in its plumage in different specimens. Whether these differences are owing merely to sex and the different periods of growth, or whether they constitute so many permanent varieties does not at present seem distinctly ascertained. Among the most remarkable are the following.

CRIMSON-TAILED COCKATOO.

This differs in having the tail of a bright crimson or scarlet colour, without any black bars, the base and tip alone being of that colour.

FUNEREAL COCKATOO.

This is described in the sixth volume of the Naturalist's Miscellany, and differs in having the middle region of the tail of a pale yellow-brown or buff-colour, varied or freckled with innumerable black specks.

BROAD-CRESTED COCKATOO.

Psittacus cristatus. *P. albus, crista plicatili, maris subtus rubra.*
White Cockatoo, with expansile crest, that of the male red beneath.

Psittacus rosaceus. } *Lath. ind. orn.*
Psittacus cristatus.
Greater Cockatoo. *Edw. pl.* 160.
Great White Cockatoo. } *Lath. syn.*
Great Red-crested Cockatoo.
Kakatoës, &c. *Pl. Enl.* 263. 498.
White-crested Parrot of Aldrovandus. *Will. orn.*
Psittacus cristatus. *Lin. Syst. Nat.*

The size of the present elegant species is that of a common fowl, and its colour white, with an extremely slight tinge of rose-colour on the head and breast, and of yellow on the inner wing-coverts and tail-feathers: on the head is a very ample crest, consisting of large and long feathers of different extent, and arching over the whole head, and which the bird can either raise or depress at pleasure: these feathers are white above, but of a fine red or scarlet beneath: the habit of the bird is stout, the tail short in proportion, and even at the end: the bill very large, strong, and of a blueish black: the orbits of the eyes bare, and of a pale ash-colour, and the legs deep cinereous. It is of a mild and docile disposition, but can rarely be taught to articulate any other word than its own name, which it pronounces with great distinctness.

M. Griffith sculp.

BROAD CRESTED COCKATOO.

1811, Aug.t 1. London, Published by G. Kearsley, Fleet Street.

The female seems to be described and figured by authors under the name of the Great White, or White-crested Cockatoo, and is of similar size and appearance with the preceding, but differs in the colour of the crest, which is entirely white.

GREAT SULPHUR-CRESTED COCKATOO.

Psittacus galeritus. *P. albus, crista plicatili elongato-acuminata sulphurea.*

White Cockatoo, with expansile lengthened-acuminate sulphur-coloured crest.

Psittacus galeritus. *P. brach: albus, crista plicatili acuminata elongata basique caudæ sulphureis. Lath. ind. orn.*

Crested Cockatoo. *White's Journ. p. 237.*

This is somewhat larger than the Red or Broad-crested Cockatoo, being of the size of a domestic cock, and measuring somewhat more than two feet in length: its colour is white, with a slight cast of yellow on the sides of the tail, and about the wing-coverts: the head is furnished with a large lengthened and pointed crest, of a fine sulphur-colour, slightly reversed at the tip: the bill is black; the legs dusky, and the tail longer in proportion than in the Broad-crested Cockatoo. It is a native of New Holland.

VAR. ?

SMALLER SULPHUR-CRESTED COCKATOO.

Psittacus sulphureus. *P. albus, crista plicatili acuminata et macula infra oculos sulphureis. Lath. ind. orn.*

White Cockatoo, with pointed sulphur-coloured crest, and a sulphur-coloured spot beneath each eye.

Psittacus albus galeritus. *Frisch. t.* 50.

Avis Kakatoeha orientalis, &c. *Seb.* 1. *t.* 59. *f.* 1.

Kakatoës a huppe jaune. *Buff. ois. Pl. Enl.* 14.

Lesser White Cockatoo. *Edw. pl.* 317.

THIS, in almost every respect except in size, is so perfectly similar to the preceding, as to make it doubtful whether it should not rather be considered as a variety than a distinct species. It has long ago been described and excellently figured in the work of Edwards. Its size is a trifle smaller than that of the common grey Parrot, and its length about fifteen inches: its colour white, with a cast of pale yellow on the breast, the sides, and the inner wing-coverts: there is also a large yellow spot beneath each eye: the crest is shaped as in the preceding bird, and is of a fine sulphur-yellow; the orbits of the eyes bare and ash-coloured; the bill black, and the legs deep lead-colour. It is a native of the Molucca islands.

M. Griffith sculp.

SMALLER SULPHUR-CRESTED COCKATOO.

1811 Aug.t 1 London, Published by G. Kearsley, Fleet Street.

RED-VENTED COCKATOO.

Psittacus Philippinarum. *P. albus, crista plicatili elongato-acuminata, crisso rubro.*
White Cockatoo, with expansile lengthened-acuminate crest, and red vent.
Psittacus Philippinarum. Red-Vented Cockatoo. *Lath.*
Petit Kakatöes à bec coleur de chair. *Buff. ois.*
Petit Kakatöes des Philippines. *Pl. Enl.* 191.
Red-Vented Cockatoo. *Brown's Illustr. pl.* 5.

THIS is the smallest of the White Cockatoos, measuring between thirteen and fourteen inches in length, and is of a white colour, with a smaller crest in proportion than the rest of the tribe, tinged with yellow at the upper part of the base, and with red beneath: the inner tail-feathers are sulphur-coloured towards the base: the bill is of a pale flesh-colour, and the legs cinereous. It is a native of the Philippine isles.

CROWNED COCKATOO.

Psittacus coronatus. *P. brachyurus viridis, crista plicatili rubra apice cærulea. Lin. Syst. Nat.*
Green Cockatoo, with expansile red crest tipped with blue.
Crowned Cockatoo. *Lath. syn.*

THIS species seems to be at present unknown in European collections: according to Linnæus it is of a green colour, and of the size of the Psittacus cristatus: the forehead yellow, the crest scarlet with bright-blue tips: the two exterior tail-feathers blue on the inner webs; and the vent red, with the margins of the feathers blue. Linnæus refers to no author for this species, but sets it down as a native of Surinam. I have sometimes been inclined to suspect that the Clusian or Ruff-Necked Parrot is, in reality, the same bird with this supposed Cockatoo.

CLUSIAN PARROT.

Psittacus Clusii. *P. viridis, capite colloque ferrugineis cæruleo undulatis, remigibus rectribusque cæruleo marginatis.*

Green Parrot, with ferruginous head and neck waved with blue, and quill and tail-feathers edged with blue.

Psittacus accipitrinus. *P. subbrachyurus viridis, capite griseo, collo pectoreque subviolaceo vario, remigibus rectricibusque upice cæruleis. Lin. Syst. Nat.. Lath. ind. orn.*

Psittacus elegans Clusii. *Ray. syn.*

Hawk-headed Parrot. *Edw. pl.* 165. *Lath. syn.*

Perroquet varié; & P. maillé. *Buff. ois. Pl. Enl.* 526.

Psittacus violaceus. Ruff-necked Parrakeet. *Lath.*

THIS singular and beautiful species seems to have been first described by Clusius, from a painting sent him in the year 1605 by Jean Plateau, who had kept the bird two years. Its size, says Clusius, was that of a Pigeon, and its head like that of a Hawk in colour: the eyes brilliant, and the neck covered with parti-coloured feathers, which, when irritated, it would set up in such a manner as to appear crested: these feathers were of a reddish colour, and on the verge of a most elegant blue: the feathers of the belly were nearly of the same colour, but with dusky clouds or variegations: the back and tail green, and the larger quill-feathers blueish.

Edwards describes this bird as of the size of a small Pigeon, with the bill, cere, and orbits blackish; the irides deep yellow or hazel; the head and throat brown, each feather marked by a longitudinal pale stripe: the neck, breast, and belly purple, the feathers edged with blue; the back, rump, scapulars, upper part of the wings, and tail fine green; the greater wing-feathers and the side-feathers of the tail dark blue towards the tips; the legs dark lead-colour. He adds, that the bird, when provoked, sets up the neck-feathers in the manner of a ruff.

The bird described by Buffon under the name of Papegai maillé is no other than a variety, differing in the brighter colour of the neck-feathers, which are rather purple-ferruginous than brown, but edged with blue as in the former. It is said to have been originally brought from India, but to have become in a manner naturalized in Guiana, where it is at present found.

It is possible that the Psittacus coronatus of Linnæus may in reality be no other than the present species; the slight differences in the colour of the forehead, &c. in the different descriptions are not such as to have much weight. In the Natural History of Guiana it is termed a Cockatoo; and as the author of that work never pretended to give a deeply scientific description of the several birds and other subjects introduced into his publication, but merely a short general notice of their most remarkable characters, he would, on seeing this bird with the neck-feathers erected, as in the representation given by Clusius, very naturally term it a Cockatoo; and for a similar reason Linnæus might have considered it as belonging to that division of the Parrot tribe.

RED AND WHITE PARROT.

Psittacus erythroleucus. *P. cinereus, uropygio remigibus rectricibusque coccineis. Lin. Syst. Nat.*

Cinereous Parrot, with bright-red quill and tail-feathers.

Red and white Parrot. *Lath. syn.*

Le Kakatöes a ailes et queue rouges. *Buff. ois.*

First described by Aldrovandus, who informs us that it is of the size of a moderately large Capon, and measures seventeen inches in length; the head and neck thick; the whole body white, but moderately shaded with dusky, so as to appear ash-coloured; the bill black, with the hooked part longer than in other Parrots; the hinder part of the back, the rump, the whole tail, and prime feathers of the wings scarlet; the legs blackish. Aldrovandus adds that in size of body it is not inferior to the largest and less vocal Parrots called Maccaws, appearing smaller only in consequence of its shorter tail.

This species seems since the days of Aldrovandus to have been very rarely imported into Europe, since the generality of naturalists content themselves with repeating the Aldrovandine description. Our countryman Dr. Hill, whose works on Natural History are not often quoted, assures us that the then Lord Petre had a beautiful bird of this species, which lived several years, and spoke very articulately. Its size was that of a well-grown Pullet; the bill remarkably large, very hooked, and black: the whole bird of a very pale and beautiful grey, having nothing of the dusky lead-colour or blueish tinge of the Common Grey Parrot, but a silvery grey, almost white; the hinder part of the back and rump of a bright and beautiful scarlet, as were also the larger wing-feathers, making an elegant variegation in the colour of the bird when sitting, but still more so when the wings were in an expanded state: the legs and feet were lead-coloured; the tail very short, hardly reaching beyond the tip of the wings, and of the same colour with the body. Dr. Hill adds that it is frequent in some of the American islands, and is also sometimes brought from Madagascar, but this, perhaps, may be doubted.

CINEREOUS PARROT.

Psittacus cinereus. *P. corpore toto cærulescente-cinereo.*
Parrot with the whole plumage of a blueish ash-colour.
Psittacus Brasiliensis cinereus. *Briss. orn.*
Maracana. *Raii syn.*
Cinereous Parrot. *Lath. syn.*

This, which is mentioned by Marcgrave in his History of Brasil, is said to resemble the Grey Parrot, but to be larger, and entirely of a blueish grey. Marcgrave supposes it to be a native of Brasil, but it seems more probable that it was imported into that part of America from Guinea.

ASH-COLOURED PARROT.

Psittacus Erithacus. *P. canus, temporibus nudis albis, cauda coccinea. Lyn. Syst. Nat.*
Grey Parrot, with naked white orbits and bright-red tail.
Psittacus cinereus seu subcæruleus. *Aldr. orn.*
Ash-coloured Parrot. *Will. orn.*
Perroquet cendrè ou Jaco. *Buff. ois. Pl. Enl.* 311. *Levaill. pl.* 99, 100, 101, 102, 103.
Common Grey Parrot.

This well known species is remarkable for its docility, the distinctness of its articulation, and its unrestrained loquacity; readily imitating every sound within its hearing. Its general size is that of a small Pigeon, and its length about twelve inches: its colour an elegant ash-grey, deeper on the upper parts, and more inclining to white beneath; the whole, though of a sober colour, having a very elegant aspect, from the deeper and lighter undulations formed by the edges of the feathers, as well as from a kind of efflorescence resembling fine powder, which, in a healthy state, is perpetually diffused over the plumage: the whole tail is of the brightest crimson; the temples or orbits of the eyes bare and white: the bill black, and the legs cinereous. This appears to have been one of the earliest imported species from the African regions, in many of which it is common. It is said to be generally brought from Congo and Angola. Buffon mentions a variety with the wings marked with red; and in the work of Edwards we meet with a mixed or pied variety, the grey ground-colour of the bird being on all parts varied with red.

The Grey Parrot seems to have been the species to which the memorable anecdote related by Gesner, and so often repeated by succeeding writers, refers. The bird belonged to King Henry the eighth, and accidentally fell out of the palace at Westminster into the Thames; and then, says Willughby, "very seasonably remembering the words it had often heard some, whether in danger or in jest use, cried out amain *A boat, a boat, for twenty pounds*. A certain experienced boatman made thither presently, took up the bird, and restored it to the King, to whom he knew it belonged, hoping for as great a reward as the bird had promised. The King agreed with the boatman that he should have as the bird, being asked anew, should say; and the bird answers *Give the knave a groat.*"

It is also probable that the Parrot mentioned by Cælius Rhodoginus, and which belonged to Cardinal Ascanius, was of this species. "I cannot," says this author, "omit an extraordinary wonder seen in our times. This was a Parrot at Rome, belonging to Cardinal Ascanius *, who purchased it for an hundred gold pieces, and which, in the most articulate and uninterrupted manner, recited the Apostles Creed as well as the best reader could have done, and which, as a most extraordinary and wonderful thing, I could not pass unnoticed."

Buffon informs us that the Grey Parrot has been known to breed in France. A gentleman at Marmande in that country had a pair which for five or six years together produced young ones, which were successfully reared: these Parrots made their nest in spring, in a cask filled with saw-dust: the number of eggs was four, one of which was always unproductive. A similar instance occurred long before at Paris according to Labat.

The Grey Parrot, like many others, is extremely long-lived, and we are sometimes told of individuals which have attained to the age of fifty, sixty, or even an hundred years.

Monsr. Levaillant gives us a curious and interesting account of one at Amsterdam which

* Probably about the year 1500.

lived in the family of a Mr. Meninck-Huysen in that city for the space of thirty-two years, and which had before lived forty-one years with Mr. Meninck-Huysen's uncle, who bequeathed it to him: thus it had lived in a state of domesticity no less than seventy-three years, and there can be little doubt that it must have been at least two or three years old at the time of its arrival in Europe. This bird, at the time Monsr. Levaillant saw it, was in a state of entire decrepitude, and, having lost its sight and memory, was in a kind of lethargic condition, and was fed at intervals with biscuit dipped in Madeira wine. In the time of its vigor and earlier age it had been a kind of Cicero among Parrots; speaking with the greatest distinctness, repeating innumerable sentences, and of so docile and obedient a disposition as always to fetch regularly its master's slippers when he required it, as well to call the servant, &c. At the age of sixty its memory began to fail, and instead of acquiring any new phrases, it began very perceptibly to lose those it had before attained, as well as to intermix in a discordant manner the words of its former language. It moulted regularly every year till the age of sixty-five, when its moulting grew very irregular, and the red feathers of its tail were supplied by yellow ones, after which no other change of plumage took place.

RED-SIDED PARROT.

Psittacus lateralis. *P. viridis, lateribus corporis tectricibusque alarum inferioribus rubris, remigibus margine cæruleis.*

Green Parrot, with the sides of the body and under wing-coverts red, and the quill-feathers edged with blue.

Green and Red Chinese Parrot. *Edw. pl.* 231.

Psittacus Sinensis. *Lath. ind. orn.*

Perroquet vert de la Chine. *Buff. ois. Pl. Enl.* 514.

Le Perroquet à flancs rouges. *Levaill. pl.* 132.

This fine species was first described and figured by Edwards, under the title of the Green and Red Parrot from China. It is however in reality a native of the Molucca islands, where it was observed by Sonnerat. It is one of the largest of the Parrot tribe, being of the size of a middling Hen. The colour of the plumage is a fine bright green, with a cast of yellow on the under parts: the sides of the body beneath the wings, as well as the whole under wing-coverts, are red: the ridges of the shoulders, the larger wing-coverts, and the outsides of the larger quill-feathers are blue: the insides of the quill-feathers and the under side of the tail are of a dusky colour, the tips of the tail on the under side inclining to yellow: the irides of the eyes are orange-yellow, and the eyes are closely surrounded with feathers: the upper mandible is orange-red, and the lower black; the legs and feet black. Monsr. Levaillant, who has given a figure of this bird, observes that authors have unnecessarily divided it into two species;

and that the *Grand Peroquet verd de la Nouvelle Guinée* of Sonnerat is undoubtedly the same with the *Green and Red Chinese Parrot* of Edwards. Dr. Latham, in his ornithology, has separated Sonnerat's bird from Edwards's, under the name of the *New-Guinea Green Parrot.* It should seem however from Sonnerat's description that the specimen observed by himself was of smaller size than that described by Edwards, since he compares its magnitude to that of an Amazon Parrot. The figure given by Buffon in the Planches Enluminées is justly censured by Levaillant, but that of Edwards, which was copied from the living bird, is extremely good.

ALDROVANDINE PARROT.

Psittacus poikilorynchos. *P. viridis, vertice flavo, macula alari rubra, rostro vario.*

Green Parrot, with yellow crown, red wing-spot, and variegated bill.

Psittacus poikilorynchos. *Aldr. orn.*

Psittacus Amazonicus? *Lyn. Syst. Nat.*

Yellow-Headed Amazons Parrot. *Lath. syn.*

This species, if such it really be (for it is generally considered as a large variety of the Amazon Parrot), seems unknown to modern ornithologists, and rests entirely on the faith of Aldrovandus, by whom it was first described.

It may certainly, says Aldrovandus, be considered as the chief of the middle-sized Parrots, being not inferior to a Hen in size, and measuring two spans from the tip of the bill to the end of the tail: the upper mandible of the bill was greenish-blue with ochre-coloured sides, and marked across the tip by a white bar: the lower mandible was lead-coloured, but yellowish on the middle: the crown of the head was adorned with gold-yellow feathers: the remainder of the plumage was green; deepest on the back, and lightest on the breast and belly, where it had a cast of yellow, but the bottoms or bases of the feathers were cinereous: the quill-feathers on their outsides towards the belly were first green, gradually changing to a blueish, amethystine, or rather purplish colour, and the remainder black: the longest of the secondaries were entirely luteous, those about the middle of the wings being on their outsides, towards the belly, first green, then dark red, then again green, and lastly, at the tips, partly violaceous, and partly black: the tail consisted of twelve feathers, of which the four lateral ones on each side were at their origin or base green on the outside, luteous on the inside, then red, and then again green with yellow tips: the remaining four, or middle feathers, were entirely green, but somewhat yellowish at the ends: the feet were lead-coloured, with two toes before and two behind, as in Woodpeckers, the inner toes being considerably shorter than the outer: the claws were crooked, as in birds of prey; the legs very short, measuring only an inch from the knees to the feet, but of considerable thickness.

It is probable that the above-described bird was in reality no other than a large yellow-headed

AMAZON PARROT.

1811, Aug. 5. London. Published by G. Kearsley, Fleet Street, & the other Proprietors

Amazon Parrot, of which the specimens, in a domestic state, are known to vary not only in the colour of the bill but in extent of plumage.

AMAZON PARROT.

Psittacus Amazonicus. *P. viridis, margine humerali maculaque alari rubris, vertice flavo, fascia frontali cærulea.*

Green Parrot, with the edges of the shoulders red, a red patch on the wings, yellow crown, and blue frontal bar.

Psittacus æstivus. *P. brachyurus viridis, luteo submaculatus, fronte cærulea, humeris sanguineis, orbitis incarnatis. Lin. Syst. Nat.*

Psittacus Amazonicus? *Lin. Syst. Nat.*

Common Amazons Parrot. *Lath. syn.*

Jamaica Parrot.
Main Parrot.
Blue-Topped Parrot.
West-Indian Green Parrot.
Brasilian Yellow-Fronted Parrot?
Yellow-Shouldered Parrot.
Yellow-Winged Parrot.
} *Lath. syn.*

Common Green Parrot, &c. *Will. orn.*

Le Parroquet Amazone. *Briss. orn. Pl. Enl.* 547.

L'Amazone. *Levaill. pl.* 84, 85, 86, 87, 88, 89.

Le Perroquet à Epaulettes jaunes. *Levaill. pl.* 98?

THE Parrots which bear the title of Amazon Parrots are natives of many parts of South America, and particularly of the regions on each side of a great extent of the river *Amazons:* some varieties are also found in the West-Indian islands. In a domestic state they are observed to vary so much in plumage as to bear many distinct names according to the particular disposition of their colours.

The general length of the Amazon Parrots is about fourteen inches: the bill in some kinds is blackish or dusky; in others pale or whitish; and in some ash-coloured, with a reddish or yellowish cast towards the base: the irides yellow or orange: the plumage bright-green, with the feathers marked by dusky or blackish margins: the tops or edges of the shoulders and a conspicuous patch on the middle of the wings bright red: the red wing-patch is usually accompanied or bounded by shades of blue, green, and yellow, which colours are completely visible only in the expanded state of the wings: the greater quill-feathers are shaded or edged with deep blue, and have dusky tips: the tail-feathers are green, but, when expanded, appear red beneath towards the base: immediately above the bill, from eye to eye, generally passes a pale but bright blue band, often accompanied on its upper part by a whitish mixture, beyond which the feathers of the crown, cheeks, and throat are of a jonquil yellow, but paler or deeper in different individuals: the legs and feet are generally either dusky or of a pale greyish brown.

The colours in these birds which seem least liable to alteration are the red and blue of the wings, and particularly the red; which remains throughout all the numerous varieties to which the species appears subject, and which have given rise to a very unnecessary multiplication of supposed species in the works of various authors.

I shall beg leave in the present instance to pursue, with some variation, the plan of the celebrated Monsr. Levaillant in the arrangement of these birds, and to consider as the genuine representative of the species the kind which he has himself described under the name of *L'Amazone.*

From a collation of six specimens, killed in their native woods, it appears that the Amazon Parrot has the front girt immediately above the bill by a blue band passing from eye to eye; and immediately above this are a few white feathers, forming a slight transverse white bar above the blue one, but not extending to each eye like the former: the remainder of the front, to some little distance beyond the eyes, is of a jonquil yellow, as are also the chin and the throat: the remainder of the plumage is of a bright and beautiful green, deepest on the back and wings, and lighter beneath, where it has a slight inclination to a blueish or glaucous colour on the breast, and to yellowish on the abdomen; and round the bottom of the thighs this latter colour forms a yellowish garter: the smallest of the wing-coverts, forming the ridge of the shoulders, for about the length of two inches and the breadth of half an inch, are of a bright and beautiful red: the larger wing or quill-feathers are externally of a deep blue with a cast of violet; the middle ones of the same colour at their tips, but red on their outward edges, thus forming a lengthened red band on the middle of the wings, but which often becomes nearly imperceptible when the wings are completely closed: the tail is deep green above, and yellowish beneath, and has some red on the upper part of each feather, which however is not seen when the tail is closed: the bill is black-brown, and the legs whitish grey: the irides of a yellow colour, more or less deep in different individuals. The total length of the male bird is fourteen inches, the tail measuring five inches.

The Female differs in having only a yellow spot on the top of the head, and in having no red on the shoulders: it has however a red bar on the middle of the wings, as in the male: the plumage of the upper parts of the body, as well as on the neck and head, has a glaucous hue, as if slightly powdered; the throat is yellowish, and the under parts dullish or subolivaceous green; the upper part of the tail deep green, growing yellowish at the edges and tip; the under part pale greenish, but with a slight appearance of red between the webs of the feathers, which however is only perceptible when the tail is spread: the ends of the larger quill-feathers are blue; the bill pale grey, and the legs and feet pale brown.

The Amazon Parrot is found in a large extent of South America, abounding in Guiana, and especially Surinam, where it causes great devastation in the plantations. It builds in the midst of impenetrable forests, the female laying four white eggs in the hollow of a tree. The young are at first quite naked, then covered with a whitish-grey down, which is gradually succeeded by the plumage. At first coming from the nest the Males and Females resemble each other; are not furnished with the red on the shoulders, and have only a small spot of yellow on the forehead: in short, in this state the young have the colours of the adult female, with this difference, viz. that the green of the plumage is of a yellower cast, and powdered above with a kind of meal, which has often caused this species in a young state to be confounded with the young of the Mealy Parrot.

VAR. 1.

YELLOW-HEADED AMAZON.

L'Amazone à tete jaune. *Buff. ois.* *Levaill. pl.* 86.
Yellow-Headed Amazons Parrot. *Lath. syn.*

In this the whole forepart of the head, to some distance beyond the eyes, is yellow, without any band of blue over the bill: in other particulars it resembles the first-described bird, except that the bill is pale, and the legs grey. This, which Buffon, after the example of Brisson, makes the chief or leading representative of the Amazon Parrot, is no other than a mere variety. It differs indeed in having a pale or yellowish bill, and legs of the same colour, and in having no other colour but yellow on the head; but Monsr. Levaillant assures us that this is no other than the gradual and accidental variation of colour: the individual which he kept for three years was in the first year tinged with blue round the front; but in each succeeding moult the blue decreased, and at length vanished altogether: the bill and legs also, which were at first brown, became yellowish-white; so that the bird appeared to be evidently a variety of the first-described kind.

VAR. 2.

BLUE-FRONTED AMAZON.

L'Amazone à calotte bleue. *Levaill. pl.* 87.

This resembles the Yellow-Headed Amazon, but has the front, immediately above the upper mandible, and rather more than half way to the eyes, blue: bill and legs dusky.

VAR. 3.

RED AND GREEN AMAZON.

L'Amazone tapirè en rouge. *Levaill. pl.* 88.

In this variety the plumage on the upper part of the head, back, wing-coverts, and breast is scalloped with red: a narrow red band passes over the base of the upper mandible, and the rump is of the same colour: the bill and legs are dusky.

VAR. 4.

PARTICOLOURED AMAZON.

L'Amazone tapirè en jaune. *Levaill. pl.* 89.

This has a richer appearance than the preceding: the forehead and cheeks are pale blue; the throat and fore part of the neck yellow; the shoulders varied with red and yellow; the larger coverts and sides of the body with green and yellow: the wings and tail as usual; the bill dusky, and the legs brown.

VAR. 5.

JONQUIL-HEADED AMAZON.

In this the whole head and neck are yellow, the shoulders varied with red and yellow, the thighs red; the bill and legs pale. A specimen of this variety is in the British Museum.

VAR. 6.

YELLOW-SHOULDERED AMAZON.

Le Perroquet a epaulettes jaunes. *Levaill. pl.* 98.
Yellow-Winged Parrot. *Lath. syn.*

In this variety the whole forepart of the head, to a considerable distance beyond the eyes, together with the whole throat, are yellow; the shoulders and thighs are also of the same colour: the wings and tail as in the rest: the bill pale, and the legs dusky. The shoulders in this variety are sometimes tinged with red, and the front of the head, immediately above the bill, with white. Monsr. Levaillant, who considers this as a distinct species, observes that in a state of domesticity, it is subject to be varied in all parts with yellow.

VAR. 7.

WHITE-FACED AMAZON.

This is mentioned by Monsr. Levaillant. It has the face all round the bill, to some distance, white, and may by that circumstance be always distinguished from the Psittacus leucocephalus, in which the upper part of the front alone is white.

There can be little doubt that many other kinds of Amazon Parrots have been described in the works of naturalists as constituting so many distinct species, which a more correct and enlarged examination would reduce to varieties; some owing to a difference of sex, and others to mere accidental alterations of colour in particular individuals. It is certain that the bill in these birds varies from a dark or blackish horn-colour to a dull or yellowish white; while the red on the shoulders is sometimes mixed with yellow, and sometimes entirely changed to that colour: nay the whole bird occasionally varies into yellow; and Monsr. Levaillant even goes so far as to maintain that there is no kind of red or green Parrot that is not liable to vary into the above-mentioned colour. There is therefore reason for supposing that even the *Psittacus Paradisi* of Linnæus may be nothing more than a yellow variety, with the major part of the plumage fringed or edged with orange-colour or red; and the *Psittacus Aurora* may be con-

PARADISE PARROT.

1811, Aug. 3. London, Published by G. Kearsley, Fleet Street.

sidered as a farther variety, still maintaining the characteristic mark of the red patch on the middle of the wings, which seems to remain through all the changes of plumage.

VAR? 8.

PARADISE PARROT.

Psittacus Paradisi. *P. flavus, marginibus pennarum rubro-aurantiis.*
Yellow Parrot, with the feathers edged with orange-red.
Psittacus Paradisi? *P. brachyurus luteus, gula ventre rectricumque basi rubris. Lin. Syst. Nat.*
Le Papagai de Paradis. *Buff. ois.*
Perroquet de Cuba. *Pl. Enl.* 336.
Paradise Parrot. *Lath. syn.*
Perroquet jaune ecaillè de rouge. *Levaill. pl.* 137.

Size of the preceding kinds: colour fine yellow, with the whole plumage edged or scalloped with red: forehead and larger wing-feathers pale grey: on the middle of the wings a red patch: bill white; legs and feet pale brown. The specimen figured in the Planches Enluminées has the whole plumage pretty strongly edged with red: in that represented in the work of Monsr. Levaillant the edging or scalloping is narrower and of a more delicate appearance.

VAR? 9.

AURORA PARROT.

Psittacus Aurora. *P. flavus, margine humerali maculaque alari rubris.*

Yellow Parrot, with the edges of the shoulders red, and a red patch on the wings.

Psittacus Aurora. *P. flavus, axillis lateribusque extimis alarum caudæque rubris. Lath. ind. orn.*

Psittacus Aurora. *P. flavus, alarum marginibus, remigibusque majoribus extrorsum in medio rubris. Lin. Mantiss.* 1771.

L'Amazone jaune. *Buff. ois. Levaill. pl.* 90.

Le Perroquet jaune. *Pl. Enl.* 13.

Aurora Parrot. *Lath. syn.*

Size of the former kinds: colour fine yellow, the shoulders slightly speckled or tinged with red, and a red wing-patch as in the rest of the tribe: tail slightly tinged with red on the under edge of some of the feathers: tips of the wings whitish: bill and legs pale flesh-colour. In a specimen figured in the work of Monsr. Levaillant the forehead is of a pale blueish ash-colour, the middle of the wing red, and the edges of the larger quill-feathers greyish; the legs and feet pale grey. According to Monsr. Levaillant this bird is no other than the Amazon Parrot in a state of extreme old age, the yellow in these birds increasing more and more as the age of the bird advances.

After having thus enumerated the principal varieties of the Amazon Parrot, I shall add a curious account of the successful raising of some young Amazon Parrots at Rome in the year 1801. The account is given in the publication entitled *Magazin Encyclopedique*, vol. 4.

In the year 1789 Monsr. Passeri purchased a pair of Amazon Parrots, the female of which, in the year 1793, laid two eggs, and in the three or four succeeding years regularly laid two or three eggs, which she sat on assiduously, being regularly attended and fed by the male: the eggs however were unproductive. At length, in the year 1800, Monsr. Passeri retired to Rome, where the Parrots were kept in a warm room, with a southern aspect. In the month of June the female again laid two eggs, which she sat on for the space of forty-one days. On the fifteenth of July one of the eggs hatched; but the young bird died the next morning, and the female forsook the remaining egg. In the spring of the year 1801 the female laid three eggs, which she diligently sat on; and in the evening of the twenty-fourth of June one of the eggs hatched; after which the male frequently brought food to the female, which immediately divided it with her young, which continued many days unfeathered and with its eyes closed. On the twelfth of July it screamed, and the male immediately brought it food. On the fourteenth it opened its eyes, and its feathers began to appear: on the sixteenth it was covered with plumes, which were fully expanded by the twenty-second: on the twenty-fourth the green quill-feathers were conspicuous, and after forty days the bird was in complete plumage, and the parent no longer regularly attended it. This young Parrot was of a bright-green colour; the wings mixed with red, blue, and violet feathers, and the tips of those near the neck yellow and red: those on the head green, with a yellow stripe towards the bill: those on the thighs yellow. After this period the male and female sometimes fed it alternately, though their regular attendance had ceased.

I cannot dismiss the history of the Amazon Parrot without observing that it appears to yield to no other kind in docility, memory, and powers of articulation; and to this species, in all probability, belonged the marvellous Parrot so solemnly commemorated by Sir William Temple, and quoted by Locke, in his celebrated work on the Human Understanding.

"I had a mind to know (says Sir William Temple), from Prince Maurice's own mouth the account of a common, but much credited story, that I had heard so often from many others, of an old Parrot he had in Brasil during his government there, that spoke, and asked, and answered common questions like a reasonable creature; so that those of his train there generally concluded it to be witchery or possession; and one of his chaplains, who had lived long afterwards in Holland, would never from that time endure a Parrot, but said they all had a Devil in them. I had heard many particulars of this story, and assevered by people hard to be discredited, which made me ask Prince Maurice what there was of truth in it. He said, with his usual plainness and dryness in talk, there was something true, but a great deal false, of what had been reported. I desired to know of him what there was of the first; he told me short and coldly, that he had heard of such an old Parrot when he came to Brasil, and though he believed nothing of it, and 'twas a good way off, yet he had so much curiosity as to send for it; that 'twas a very great and a very old one; and when it came first into the room where the Prince was, with a great many Dutchmen about him, it said presently, *What a company of white men are here!* They asked it what he thought that man was; pointing at the Prince. It answered *Some General or other.* When they brought it close to him, he asked it, *D'ou venes vous?* It answered *De Marinnan.* The Prince. *A qui estes vous?* The Parrot. *A un Portugais.* Prince. *Que fais tula?* Parrot. *Je garde les poulles.* The Prince laughed and said, *Vous gardez les poulles?* The Parrot answered, *Ouy, moy & je scay bien faire;* and made the chuck four or five times that people use to make to chickens when they call them." "I set down," adds Sir William Temple, "the words of this worthy dialogue in French just as Prince Maurice said them to me. I asked him, in what language the Parrot spoke, and he said in Brasilian. I asked whether he understood Brasilian. He said no, but he had taken care to have two interpreters by him, the one a Dutchman that spoke Brasilian, and the other a Brasilian that spoke Dutch: that he asked them separately and privately, and both of them agreed in telling him just the same thing that the Parrot said."

In modern times we have heard of no bird of this genus that could rival the celebrated Parrot so many years in the possession of the late Colonel O'Kelly. The following particulars, politely communicated by Sir Henry Englefield, Bart. afford a convincing proof of the memory and docility sometimes exemplified in the present race of Parrots.

"Col. O'Kelly's Parrot was of a beautiful green colour, and had, I think, a spot of red on each shoulder. I saw him in company with eight or ten other strangers. At first the bird seemed alarmed at the number of people, and was quite sullen. The housekeeper was sent for, and she gradually brought him into good humour by caressing him, talking, and singing to him. The first tune he sung was *The banks of the Dee*. He sung several verses of this by snatches, pausing between whiles, as if his memory failed him. The maid then helped him, by going on where he left off, and then he took her up, not repeating what she had sung, but going on from her last words. He pronounced the words of the song very distinctly, much more so indeed than when he spoke, for his articulation in speaking was not remarkably clear. The tone of his singing was very odd. It seemed like an automaton imitating a human voice; but the intonation was very good, and had he been taught by a good singer instead of a kitchen-maid, I have no doubt that he would have sung very agreeably. The maid prompted him to sing *God save the King.* He sang all the verses of it, but every now and then wandered back into the *Banks of the Dee*, which seemed his favourite. Afterwards he sung the *hundredth Psalm*, and one or two Scotch songs the names of which I forget. I think in the whole he sung six different songs, all with words to them. He did not sing them all entirely through, but seemed as if he could have done it had he been quite in good humour. Col. O'Kelly told us that he knew about twenty songs, and, from what I saw and heard, I can readily believe it. Ices and cakes were brought to the company. A teaspoon with cream ice in it was given to the bird. He took it in one foot, standing on his perch on the other, held it up to his bill, and supped it up gently, looking round and talking at intervals to the company in a most singular manner, and with an air of intelligence quite surprising. While thus employed, somebody pushed him with a finger; he turned quick round and said in an angry tone *Don't throw me down.* Col. O'Kelly told us that his power of catching sounds was quite astonishing: that when a newspaper had been read aloud in his presence, the Colonel, on coming into the room half an hour after, had, as he opened the door, been convinced by the sound that the same person was still reading aloud, and had been scarce able to believe that it was the Parrot repeating to himself inarticulate sounds precisely in the tone and manner of the reader. He also delighted in imitating the Puppet-shows that exhibit in the streets. This is all I can distinctly remember of this very extraordinary bird."

AOUROU PARROT.

Psittacus Aourou. *P. viridis vertice flavo, fascia frontali superciliisque cæruleis, genis aurantiis, macula alari rubra.*

Green Parrot, with yellow crown, blue frontlet and brows, orange cheeks, and red wing-spot.

Le Perroquet Aourou-couraou. *Levaill. pl.* 110. 111.

L'Aourou-couraou No. 5. *Buff. ois.*

Le Perroquet Amazone? *Pl. Enl.* 547.

Brasilian Yellow-Fronted Parrot. Var. E. of Common Amazon. *Lath.*

Psittacus agilis. *Lin.* Little Green Parrot. *Edw.* 168. A young bird?

This, which, from the general disposition of its colours, might be supposed no other than a variety of the Psittacus Amazonicus, is, according to Levaillant, a distinct species, differing in being of smaller size, of a duller or darker green, more especially on the upper parts, which have a greyish or brownish cast in different lights, but particularly in the bill, which is yellowish with dark brown or blackish edges and tip: across the forehead and over each eye runs an ultramarine-blue bar, and the cheeks and throat are rather orange-coloured than yellow: the disposition of the colours on the wings and tail is similar to that of the Amazon Parrot, but there is no red on the shoulders, and a narrow patch or spot of bright yellow appears on the edge of the wing towards the middle. The female is rather smaller, with the blue bar on the forehead paler, and the general colour some-

AOUROU PARROT.

1811. Aug. 1. London. Published by G. Kearsley, Fleet Street, & the other Proprietors.

what less vivid. Native of Guiana, where it is very common.

It is to this species, according to Monsr. Levaillant, that the *Psittacus agilis* of Linnæus or *Little Green Parrot*, Edwards Pl. 168. is to be referred; Monsr. Levaillant supposing it to be no other than the *Aouro-couraou* in its young state, and consequently an imaginary species. Against this supposition, however, militates the assertion of Edwards, its first describer, who says, "I kept it alive some years; it was a brisk, lively, nimble bird, and talkative in a language unknown to me." Now it seems probable that had it been, as Monsr. Levaillant supposes, a young bird, or in its first years of plumage, it would afterwards have gradually exhibited the proper colours of its species. It remains, however, to give an abridgement of the description of Edwards, and to leave the scientific ornithologist to judge for himself on the subject.

"This bird is about the size of a small Pigeon: the bill of a light ash-colour: round each eye is a small space of bare skin of a light ash-colour: the whole head, neck, and body, both above and beneath, are of a green colour, though lighter and more inclining to yellow on the throat, breast, belly, thighs and coverts under the tail: the greater quills of the wings are of a dusky or black colour, their outer webs being blue almost to their tips: the remainder of the quills next the back are green: amongst the first row of covert-feathers there is one red feather, which falls over the bottoms of the blue quills: all the remainder of the covert-feathers, both above and beneath, are green: the two middle feathers of the tail are a little longer than those on the sides, but not so as to make it one of the long or pointed-tailed Parrots: the upper side of the tail is green, except the outer webs of the two outermost feathers, which are blueish: the inner webs of the tail-feathers are red, till within less than an inch of their tips, which are wholly green; and the outer webs of the two outer feathers, which are blue above, are light green beneath: the legs and feet are ash-colour. I take it to be from the West Indies."

RED-BAR PARROT.

Psittacus signatus. *P. viridis, remigibus extus cæruleis, tectricibus majoribus rubris, cauda subtus versus basin rubra.*

Green Parrot, with the quill-feathers blue externally, the larger coverts red, and the tail red beneath towards the base.

Le Petit Perroquet Vert. *Levaill. pl.* 105.

Little Green Parrot? *Edw. pl.* 168. an var?

This species has all the upper parts of its plumage of a fine green colour, with a yellowish cast on the under parts, and a slight blueish or glaucous tinge on the back and shoulders: the largest row of wing-coverts, or those immediately covering the quill-feathers, are scarlet: the larger quill-feathers dusky, with the outward edges blue: the smaller quill-feathers green, like the back: the tail green, but with all the feathers except the two middle ones red on their inner webs from the base to two thirds of their length: the bill very pale ash-colour, and the legs and feet dull grey. According to Monsr. Levaillant's plate, this bird should seem to measure about ten inches and three quarters from the tip of the bill to the end of the tail. It is said to be a native of Brasil, and seems so very nearly allied to Edwards's Little Green Parrot, mentioned under the description of the immediately preceding species, as to make it very doubtful whether Monsr. Levaillant be correct in maintaining either Edwards's bird to be merely a young Oura-Ourao, or this to be different from the bird described by Edwards, and which, as has been before observed, would, in all probability, have gradually displayed the distinctive colours of the *Aouro-couraou*, had it been a young of that species.

MEALY PARROT.

Psittacus pulverulentus. *P. viridis albo-pulvcreus, macula verticali flava, alari rubra, remigibus apice nigro-cæruleis.*

Mealy-green Parrot, with yellow spot on the crown, red patch on the wings, and quill-feathers blackish-blue towards the tips.

Psittacus pulverulentus. Mealy Parrot. *Lath.*

Meunier, ou Crik poudrè. *Buff. ois.*

Le Perroquet Meunier. *Levaill. pl.* 92.

This species, which was first described by Buffon, seems to be next in size to the Aldrovandine

Parrot. Buffon, indeed, says it is the largest of all the American Parrots next to the Maccaw tribe; but he evidently forgets the measure of the bird above-mentioned. The Mealy Parrot, according to the admirable figure given by Monsr. Levaillant, appears to measure about fifteen inches from the tip of the bill to the end of the tail, and is consequently a size larger than the Amazon Parrot, to which it is in some points considerably allied. Its colour is a beautiful green, deepest on the head and wings, and palest on the breast and belly: the whole of the plumage, but particularly on the shoulders, upper part of the back and wings, breast, and belly, has an elegant glaucous cast, owing to the secretion of a whitish powder, which, in the healthy state of the bird, never fails to diffuse itself over the feathers, and which is observable, though in a much slighter degree, on the plumage of some other parrots: on the top of the head is a moderately-sized yellow spot: the edge or border of the wing, a little beyond the shoulders, is red for the length of about two inches: the innermost of the larger quill-feathers are red on their middles, forming a longish patch of this colour, as in the Amazon Parrot, and the remainder is deep blue, darker or blacker at the tips: the outside feather on each side the tail, which is larger in proportion than in the Amazon Parrot, is deep-blue on the outside web, growing somewhat paler at the tip: the remaining tail-feathers are green, with slightly yellowish tips, but have not the least appearance of red on their inner webs, as in the Amazon Parrots: the bill is dusky or blackish, with the upper mandible yellowish on each side towards the base; the legs and feet deep cinereous, and the irides of the eyes yellow.

The Mealy Parrot is a native of South America, and particularly of Cayenne, and, when in a young state, is often mistaken for the Amazon Parrot in its incomplete plumage, or before it has arrived at its full colours. In a domestic state it is much esteemed for its size and singularity of appearance, as well as for its mildness and docility of disposition.

DUFRESNE'S PARROT.

Psittacus Dufresnianus. *P. viridis, genis colloque antico subviolaceis, fascia frontali maculaque alari aurantiis.*

Green Parrot, with subviolaceous cheeks and fore-neck, and orange frontal band and wing-patch.

Le Perroquet Dufresne. *Levaill. pl.* 91.

THIS is considered by Monsr. Levaillant as a new and extremely rare species. Its size is nearly equal to that of the Mealy Parrot, and its colour a deep or full green: immediately above the base of the upper mandible, from eye to eye, runs an orange-yellow bar: the cheeks and sides of the neck are tinged with blue: the ridge of the wings, and the outside edges and tips of the larger quill-feathers are deep-blue; the inner webs of the latter marked in the middle by a broad orange-coloured bar: the bill is pale brown, and the legs dusky ash-colour. Native of Cayenne.

BRASILIAN GREEN PARROT.

Psittacus Brasiliensis. *P. viridis, capistro rubro, genis cæruleis, cauda apice flava.*

Green Parrot, with red frontlet, blue cheeks, and yellow-tipped tail.

Psittacus Brasiliensis. *P. brachyurus viridis, facie rubra, temporibus cæruleis, orbitis cinereis. Lin. Syst. Nat.*

Brasilian Green Parrot. *Edw. pl.* 161.

Blue-Faced Parrot. var. C. *Lath. syn.*

Perroquet à joues bleues. *Levaill. pl.* 106.

This is one of the most beautiful of the short or even-tailed Parrots, and in size is at least equal, if not rather superior to the Psittacus erithacus or Common Grey Parrot. Its plumage is fine grass-green, rather paler beneath; the feathers edged or fringed with purplish-brown: the front, all round the base of the bill, is bright-red; the cheeks fine deep-blue, and the top of the head yellow: the edge of the wing, at some distance beyond the shoulders, is red; all the wing-coverts and the shorter quill-feathers edged with yellow: the alulet pale, and the larger quill-feathers deep or dusky blue: the outside feather on each side the tail deep blue, tipped pretty deeply with yellow; the next feather red with a similar yellow tip, and all the remaining ones green with yellow

BRASILIAN GREEN PARROT.

1811. Aug. 1. London. Published by G. Kearsley, Fleet Street, & the other Proprietors.

tips: the bill pale or whitish, and the legs and feet dusky.

The Count de Buffon considers this bird as a variety of his *Crik à tete bleue*, which is the *Blue-Faced Parrot* of Edwards; but Monsr. Levaillant assures us that this is a mistake, and that the present bird is an entirely distinct species. It is a native of South America, and particularly of Brasil.

BLUE-FACED PARROT.

Psittacus cærulifrons. *P. viridis, facie remigibusque primoribus cæruleis, gula maculaque alari rubris, cauda apice flavescente.*

Green Parrot, with blue face and larger quill-feathers, red throat and wing-patch, and yellowish-tipped tail.

Blue-faced Green Parrot. *Edw. pl.* 230.

Psittacus autumnalis var. β. *Lath. ind. orn.*

Le Perroquet Bouquet. *Levaill. pl.* 135.

THIS is one of those Parrots which were first described by Edwards, and is certainly much allied to the *Psittacus Brasiliensis*. Its size, according to Edwards, is that of a small-sized Hen or Pullet: its colour is green, paler beneath, and the edges of the plumage on the wings yellowish: the whole front, to some little distance beyond the eyes, is blue, somewhat inclining to violet: the throat, immediately under the blue of the chin, and to some distance down the breast, red, terminating in a pointed manner: the larger quill-feathers blue; but three or four of the secondaries are red towards their base, thus forming a red patch on the wings: the tail is green, the feathers tinged with red at the base, and pretty deeply edged with yellowish or apple-green: the bill dusky ash-colour, the middle part of the upper mandible inclining to orange: the orbits bare and flesh-coloured, and the legs and feet of the same hue. Native, according to Buffon, of Guiana. A supposed variety of this bird is mentioned by Dr. Latham (from Buffon,) which differs in having the head varied with red and dull white.

AUTUMNAL PARROT.

Psittacus autumnalis. *P. brachyurus viridis, fronte remigumque macula coccinea, vertice remigibusque primoribus cæruleis. Lin. Syst. Nat.*

Green Parrot, with red front and wing-patch, orange cheeks, and blue crown and quill-feathers.

Autumnal Parrot. var. B. Blue-Faced Parrot. *Lath.*

Lesser Green Parrot. *Edw. pl.* 164.

Le Perroquet à joues orange. *Levaill. pl.* 111.

First described by Edwards, and allied to the Blue-Faced and Blue-Cheeked Parrots, but smaller than either, being, according to Edwards, of the size of a middling Pigeon. Its colour is dark green above, paler or inclining to apple-green beneath; the thighs and vent-feathers pale yellow-green; the forehead red, the top of the head blue, and the cheeks orange, fading into yellow as it joins the plumage on the sides of the neck: the

WHITE-FRONTED PARROT.

1811. Augt 1. London. Published by G. Kearsley, Fleet Street, & the other Proprietors.

alulet is blue; the ridge of the wing yellow, and the larger quill-feathers dusky, but blue on the outer edges and red towards the base, forming a red patch on each wing: the bill and orbits are whitish, and the legs and feet ash-colour. Supposed to be a native of Guiana.

This and the two last-described birds are, in Dr. Latham's Ornithology, considered as constituting one and the same species, under the name of the *Blue-Faced Parrot*, of which the present and the *Brasilian Green Parrot* rank as varieties. In this he follows the example of Buffon; but Edwards considered them as so many distinct species, and Monsr. Levaillant is loud in his declarations of hostility against the arrangement of the Count de Buffon.

WHITE-FRONTED PARROT.

Psittacus leucocephalus. *P. viridis, fronte alba, gula rubra, remigibus coccineis.*

Green Parrot, with white front, red throat, and blue quill-feathers.

Psittacus leucocephalus. *Lin. Syst. Nat.*

Psittacus leucocephalus. *Aldrov. orn.*

Amazone à tete blanche. *Buff. ois.*

Perroquet &c. *Pl. Enl.* 335. 549. 548.

White-Headed Parrot. *Edw. pl.* 166.

White-Fronted Parrot. *Lath. syn.*

Perroquet à face rouge. *Levaill. pl.* 107. 108. 109.

This highly elegant species is a native of the larger West Indian islands, and has by some been named the White-Headed Amazon Parrot. Its size is somewhat smaller than that of the Amazon, measuring from ten to twelve inches in length. The plumage is light green, with a yellowish cast on the under parts of the body; and the feathers are elegantly marked or fringed by a dusky or blackish edge. The front, to a greater or smaller extent in different individuals, is milk-white; in some extending half way over the head and including the eyes; in others reaching only about a third part over the head; and scarcely passing beyond the eyes: the upper part of the head, behind the white feathers, is of a pale but bright blue, and is sometimes slightly intermixed with red: the lower part of the cheeks, the throat, and under part of the neck to a considerable extent, are of a bright scarlet colour: the larger wing-feathers bright blue with dusky tips, and some of the larger coverts are often tinged with the same colour: the rest of the wing-feathers are green, and are destitute of the red bar, so conspicuous in the Amazon Parrots: the edge of the shoulder is, however, in some specimens, slightly tinged with red: the lower part of the belly is generally more or less varied or scalloped with red, but in some individuals is marked by a large bed of purplish red, or darker than that on the throat: the tail is green above, with the exterior webs of the two outside feathers blueish; but on the under part the three or four lateral feathers are red from their base to about half their length: the bill is pale or white; the orbits surrounded by a white skin, the irides hazel, and the legs dusky. Native of St. Domingo, where it is common, but said to occur also in Jamaica, &c.

The variations exhibited by this bird, (at least in the full-grown male,) are rarely such as to prevent the ready determination of the species; the bill being constant in its colour, and the white front pretty conspicuous. In some the edges of the shoulders are white, in some reddish, but in the major part green: the red on the throat is sometimes confined to a small space, and that on the belly is in some specimens of a brownish cast. Monsr. Levaillant also mentions some specimens which had their green plumage varied or scolloped with red. In the female, according to Levaillant, the plumage is of a more dusky green than in the male, and the front is red instead of white: there is also no appearance of red either on the throat, the abdomen, or beneath the tail. The female has been erroneously described by Buffon as an entirely different species, under the title of *Papegai a bandeau rouge*. The young of both sexes, at first leaving the nest, are entirely similar to each other; but after the first moulting the male exhibits the white front as well as a red spot on the throat, and beneath the tail-feathers; but the female at that period is entirely green, except on the edges of the quill-feathers, which are blue. The young male, after its moulting, is the *Psittacus albifrons* or *White-Crowned Parrot* of Dr. Latham, who has introduced it into his Ornithology on the faith of Sparmann's description given in the work entitled Museum Carlsonianum. Monsr. Virey also, in Sonnini's edition of Buffon, has continued the error, by repeating Sparmann's description of the young male under the title of *Papegay a front blanc.*

BLUE-CAPPED PARROT.

Psittacus gramineus. *P. viridis subtus olivaceo-flavescens, pileo cæruleo.*

Green Parrot, yellowish-olivaceous beneath, with blue crown.

Psittacus gramineus. Amboina Parrot. *Lath.*

Grand Perroquet vert à tete bleue. *Buff. ois.*

Perroquet d'Amboine. *Pl. Enl.* 862.

Perroquet a calotte bleue. *Levaill. pl.* 121.

This is a large and elegant species, equal in size to the Mealy Parrot, and measuring sixteen inches in length: the crown of the head, from the base of the upper mandible to some distance beyond the eyes, is sky-blue: the back of the neck and whole upper parts of the bird full grass-green; the whole under parts, from the base of the lower mandible to the end of the tail, pale yellow-green. In the description given by the Count de Buffon the under parts are said to be olive-green, and are represented of that colour in the Planches Enluminées, No. 862. but in the elegant figure given by Monsr. Levaillant the colour is rather pale yellow, accompanied by a slight subolivaceous cast: the bill is pale rufous or yellow-brown, and from its corners to the eyes runs a narrow black streak: the legs are dusky. It is supposed to be a native of Amboina, and is considered as a very rare species.

FESTIVE PARROT.

Psittacus festivus. *P. brachyurus viridis, fronte purpurascente, superciliis gulaque cœruleis, dorso sanguineo. Lin. Syst. Nat.*

Green Parrot, with purplish forehead, blue brows and throat, and red back.

Tavoua ou Tabué. *Buff. ois.*

Perroquet Tahué de Cayenne. *Pl. Enl.* 840.

Le Perroquet Tavoua. *Levaill. pl.* 129.

Festive Parrot. *Lath. syn.*

Size of the Amazon Parrot: colour deep green: over the base of the upper mandible from eye to eye a very deep crimson bar: crown of the head, brows, and cheeks tinged with blue, and on the throat a deep-blue spot: larger wing-feathers edged and tipped with deep-blue: lower part of the back, or rump scarlet: bill and legs black: irides orange. Native of Guiana. It is said by Buffon to be of an active and lively nature, and to articulate even more clearly than the Common Grey or Guinea Parrot, but to be of a capricious and malignant disposition. It is observed by Levaillant that this species is not often seen in a living state in Europe: one which he observed in a state of domesticity was variegated with red on the green part of the plumage, and the red on the back had extended itself almost to the shoulders. This change began about the fifth year of its captivity, when the wing-coverts became chequered with red, and at every successive moulting the red feathers increased in number, till at length there seemed to be as many red as green feathers on the bird. Monsr. Levaillant adds that though the individual described by Buffon was of a perverse temper, yet he has himself seen one which was remarkably tame and good-natured; and that it is therefore probable that the faults of that commemorated by the Count de Buffon were owing rather to improper education than to any inherent malignity of disposition.

SOUTHERN BROWN PARROT.

Psittacus Nestor. *P. fuscus, capite incano, collo inferiore castaneo, uropygio crissoque castaneo-rubris. Lath. ind. orn.*

Brown Parrot, with grey head, chesnut underneck, and reddish-chesnut rump and vent.

Southern Brown Parrot. *Lath. syn.*

Psittacus Australis. *Museum Leverianum. No. 2. t. 9.*

A LARGE species, measuring about sixteen inches in length: general colour ash-brown; head pale grey: under parts of the neck chesnut: lower part of the belly the same, but of a somewhat redder cast: bill large, very much hooked, and horn-coloured: legs grey. Native of New Zealand.

RED-CROWNED PARROT.

Psittacus galeatus. *P. brachyurus cristatus fusco-viridis, vertice rubro, abdomine rubro viridique undulato. Lath. ind. orn. suppl.*
Dusky-green Parrot, with red crown, and abdomen undulated with red and green.
Red-crowned Parrot. *Lath. syn. suppl.* 2.

SIZE of the Grey Parrot: length about fifteen inches: colour dusky with a green gloss, the margins of the feathers pale: under parts paler than the upper, and undulated with dull red: feathers on the head deep-red, full, and forming a kind of crest: bill yellow-brown; legs dusky. Native of New Holland.

DAMASK PARROT.

Psittacus infuscatus. *P. olivaceo-fuscus thalassino varius, margine humerali femoribusque aurantiis, rostro albido.*
Olivaceous-brown Parrot, varied with sea-green; the edges of the shoulders and thighs orange-coloured.
Le Perroquet à franges souci. *Levaill. pl.* 130. 131.

THIS species is nearly equal in size to the Common Grey Parrot, and like that bird, is of a stout habit: the colour of the head, neck, and breast is olivaceous grey-brown; the abdomen, rump, and vent-feathers, bright glossy sea-green; back and wings greenish brown; edges of the shoulders, alulets, and thighs orange or marigold-colour; bill strong and whitish; legs dusky. This bird, which is described by Levaillant, is a native of the eastern parts of Africa, inhabiting deep forests, and not advancing so far southwards as the Cape of Good Hope; being found in lat. 32 south, and consequently far beyond the limits so vainly and absurdly assigned by Buffon to the Parrot tribe. It delights particularly in the fruit of the tree called by the Dutch in those regions *Geele Hoult* (bois jaune,) or *Wild Kaersen, (Cerisier sauvage,* Fr.) Vast flocks fly into the above-mentioned Eastern forests, in which they pass all the warm season, and return again to higher latitudes on the approach of the rainy season, after they have reared their young. These flocks often fly at such a height as to be heard without being seen. During the breeding-season each pair remain inseparable, either in the hollow of a tree or of a rock, where the female lays four white eggs, about the size of Pigeon's eggs, on dry leaves, moss, or decayed wood; the male partaking the task of incubation with the female. The exact period of this could not be ascertained by Monsr. Levaillant, but he is inclined to think that the young are hatched in less than twenty-four days. The young are at first entirely naked, and are afterwards covered with greyish-white down, through which, in the space of about six days, the shafts of the feathers may be perceived; and at this period nothing can be more ugly than the young birds. In about six weeks they are entirely covered with plumage, but remain in the nest a considerable time longer, and are fed by the parents, who disgorge in the manner of Pigeons. At the age of about two months they begin to feed themselves, at first accompanying the parents in their flights, by whom they are afterwards driven away, when they form a separate party.

Every day, at the same hour, says Monsr. Levaillant, the Parrots of this species fly to the water in order to bathe themselves, in which operation they take great delight: their hours of feeding are also extremely regular, and their whole day is distributed by rule. At the dawn of morning the whole flight of each canton assemble, and alight, with much noise, on one or more dead trees, according to the size of the flock, and there, displaying their wings to the first rays of the sun, recal to the mind of the spectator the idea of some ancient race, of simple manners, assembled on some hill, in order to chant a hymn in honour of the God of Day. The reason, however, of this assembly of the Parrots is to warm and dry their plumage, moistened and chilled by the dews of night, which in these regions is often very cold, and always damp. When they are once warmed and dried, they arise in small flocks, and fly about in quest of their favourite fruit, the stone of which they break in order to obtain the kernel; in this resembling the generality of the Parrot tribe, which prefer the kernel to the fruit itself. This their morning repast continues till about ten or eleven o'clock, at which time all the separate flocks fly to the water to bathe. When the heat of day commences they again seek the recesses of the woods, in order to enjoy the refreshment of the shade; and at this time they keep a silence so profound, that not a single bird shall be heard by a person sitting beneath a tree which contains legions in its upper part; but on the report of a gun the whole group fly off with the rapidity of lightning, and with a confused mixture of the most discordant cries. When this their resting or sleeping-time is elapsed, they again disperse, in order to gain their second or evening meal, when all the flocks of the whole canton again assemble, with much noise and animation; and this is the signal for their second visit to the water, which is often at a great distance, since no other than the purest water will please them. They are then seen huddling or rolling over each other pell-mell on the banks of the water, frolicking together, dipping their heads and wings into the water in such a manner as to scatter it over all their plumage, and exhibiting a most entertaining spectacle to the observer. This ceremonial being finished, they revisit the trees on which they assembled at sunrise, where they sit in order to adjust and preen their feathers; and this being finished, they fly off in pairs, each pair seeking its particular retirement in the woods, where they wait till morning.

The male and female of this species resemble each other, except that the male is a degree larger, and the colours, especially the marigold and green, somewhat brighter than in the female.

Monsr. Levaillant describes a variety of this species, which he took from the nest, and bred in a domestic state, and which, after having lived some time in perfect health, at length began to languish, and, on moulting, exhibited the appearance of marigold-coloured feathers on the head: after this, every feather which it lost was replaced by one of a marigold-colour. This bird died on board the ship in which it was conveying to Europe. Monsr. Levaillant also once, and but once, saw an approach to this variegated appearance in the wild bird, which, however, had only a few marigold-coloured plumes on the front of the head; and here Monsr. Levaillant observes, that he once saw, among a parcel of birds sent from Cayenne, two specimens of the Parrot called *Maccavouanne (Psittacus Makavouanna,)* variegated with red, though they had been killed in their native woods: a proof that this kind of variegation in the Parrot tribe sometimes takes place in a state of nature, as well as in a state of confinement.

MASCARINE PARROT.

Psittacus Mascarinus. *P. cinereo-fuscus, facie nigra, rostro pedibusque rubentibus, cauda versus basin albida.*

Grey-brown Parrot, with black face, reddish bill and legs, and tail whitish towards the base.

Psittacus mascarinus. *P. fuscus, facie nigra, cauda albente. Lath. ind. orn. Lin. Mant.* 1771. *p.* 524.

Mascarin. *Buff. ois. Pl. Enl.* 35. *Levaill. pl.* 139.

A DEGREE larger than the Grey Parrot, of a somewhat more slender appearance, and with the tail longer in proportion: head and neck grey, but the whole front, as far as the eyes, and beneath the throat towards each side of the neck, deep black, forming the appearance of a kind of mask: remainder of the plumage deep grey-brown above, and paler beneath: tail whitish on each side towards the base. Native of Madagascar.

VASA PARROT.

Psittacus Vasa. *P. subcinereo-niger, rostro orbitisque albidis, cauda subelongata.*

Greyish-black Parrot, with whitish bill and orbits, and somewhat lengthened tail.

Le Grand Vasa. *Levaill. pl.* 81.

LENGTH twenty-one inches French from the top of the head to the end of the tail: bill large and whitish: colour of the whole bird black, with a cast of grey or brown, according to the direction of the light: orbits naked and whitish; legs black. Said to be a native of the southern parts of Africa, but not observed in those regions by Levaillant, who describes it from a specimen in the cabinet of Monsr. de Breukelvaart of Amsterdam.

SMALLER VASA PARROT.

Psittacus niger. *P. nigro-cærulescens, rostro orbitisque albidis, cauda subelongata.*

Blueish-black Parrot, with whitish bill and orbits, and somewhat lengthened tail.

Psittacus niger. *P. corpore cærulescente-nigro, rostro orbitisque albidis. Lath. ind. orn.*

Psittacus niger. *P. cauda elongata æquali, corpore cærulescente-nigro, rostro orbitisque albidis. Lin. Syst. Nat.*

Black Parrot of Madagascar. *Edw. pl.* 5.

Le Vasa ou Perroquet noir. *Buff. ois.*

Perroquet noir de Madagascar. *Pl. Enl.* 500.

Le Petit Vasa. *Levaill. pl.* 82.

This, according to Monsr. Levaillant, is by no means to be confounded with the preceding, from which it differs greatly in size, though similar in its general appearance: the bill is also small in proportion to the bird, whereas in the preceding it is very large. The Smaller Vasa measures about fourteen inches in length, and the tail, when expanded, is wider at the end than in the former species. It is a bird of a very lively disposition, and though it cannot be taught to speak, imitates with great readiness and exactness any tune which it hears, as well as the barking of dogs, mewing of cats, &c. &c. in the manner of the Mocking-Bird. It is a native of the island of Madagascar.

GREAT-BILLED PARROT.

Psittacus macrorynchos. *P. viridis, dorso alisque cærulescentibus, tectricibus nigris luteo marginatis.*

Green Parrot, with blueish back and wings, with black coverts edged with orange-colour.

Psittacus macrorynchos. Great-Billed Parrot. *Lath.*

Perroquet à bec couleur de sang. *Buff. ois. Levaill. pl.* 83.

Perroquet de la Nouvelle Guinèe. *Pl. Enl.* 713.

Length from twelve or thirteen to fifteen inches: colour fine green, with a blueish cast on the upper part of the neck and back, and a still stronger one on the wings: the smaller wing-coverts are black, beautifully edged with pale orange, the lowest row or those nearest the larger coverts being of a green colour, with paler edges inclining to yellow: the larger quill-feathers are of a greenish blue on their outsides, but blackish at their tips and on the inner webs: the bill very large, and of a bright red; the legs dusky. Native of New Guinea, and is considered as a rare species. Monsr. Levaillant justly observes that the figure given in the Planches Enluminées was copied from an ill-prepared specimen. Both the head and bill in that figure are indeed represented beyond their true proportions.

GREAT BILLED PARROT.

H. Griffith, sculp.

18[illegible], Aug.t 1. London, Published by G. Kearsley, Fleet Street.

HAVANNA PARROT.

Psittacus Havanensis. *P. viridis, subtus subviolaceus, facie remigumque marginibus cæruleis, macula alari rubra, crisso flavo.*
Green Parrot, subviolaceous beneath, with the face and edges of the quill-feathers blue, red wing-spot, and yellow vent.
Psittacus Havanensis. Havanna Parrot. *Lath.*
Crik à face bleue. *Buff. ois.*
Perroquet à face bleue. *Levaill. pl.* 122.
Perroquet de la Havane. *Pl. Enl.* 360.

Length from twelve to fifteen inches: whole face considerably beyond the eyes, sky-blue: edges of the larger quill-feathers deep blue: under parts pale lilac undulated or edged with green: on the wings a patch of red, as in the Amazon Parrot: vent yellow: bill and legs horn-colour. Native of Mexico, as well as of Cuba and some other American islands. The specimen described by Buffon was received from Havanna. This species, according to Levaillant, may be numbered among the most docile and intelligent of the whole Parrot tribe.

COCHINCHINA PARROT.

Psittacus Cochinsinensis. *P. cæruleus, fronte nucha pectore abdomineque coccineis, fascia tectricum remigibus caudaque nigris.*

Blue Parrot, with scarlet forehead, nape, breast, and belly; black bar across the coverts, and black quill-feathers and tail.

Psittacus Cochinsinensis. *P. cæruleus, fronte nucha collo inferiore pectore abdomineque medio coccineis, tectricibus alarum fascia remigibus caudaque nigris. Lath. ind. orn.*

DESCRIBED by Dr. Latham from a drawing in the possession of Lady Impey: size not mentioned. Native of Cochinchina.

PARAGUAN LORY.

Psittacus Paraguanus. *P. niger, dorso gula pectore abdomineque coccineis.*

Black Lory, with red back, throat, breast, and belly.

Psittacus Paraguanus. *P. coccineus, capite cervice crisso cauda humeris alisque nigris. Lath. ind. orn.*

Paragua. *Raii syn. Will. orn. Marcg. Bras.*

Paraguan Lory. *Lath. syn.*

THIS, which seems to have been first described by Marcgrave, is said to be of the size of the Amazon Parrot, and is supposed to be a native of Brazil.

GRAND LORY.

1811. Aug.t 1. London. Published by G. Kearsley, Fleet Street, & the other Proprietors.

GRAND LORY.

Psittacus grandis. *P. coccineus, zona corporis subviolacea, margine humerali remigumque cæruleis, cauda subtus flavescente.*

Scarlet Lory, with a subviolaceous zone round the body, edges of the shoulders and quill-feathers blue, and tail yellowish beneath.

Psittacus grandis. Grand Lory. *Lath.*

Le Grand Loris. *Buff. ois.* *Levaill. p.* 126. 127. 128.

Lory de la Nouvelle Guinèe. *Pl. Enl.* 683.

Size of the Amazon Parrot: colour deep scarlet; the upper part of the body clouded by a broad zone of pale violet, sometimes intermixed with green, and surrounding the breast and upper part of the back: edges of the shoulders bright blue; larger quill-feathers of the same colour: vent-feathers yellow; inner edges of the tail-feathers yellow towards the tip: bill black; legs dusky. Native of New Guinea and the Moluccas, and said to be sometimes found in Ceylon.

Monsr. Levaillant describes and figures a variety in which the pale violet zone is scarcely perceptible, and the feathers of the breast and belly edged with green; and another in which the breast and belly are very strongly clouded or scalloped with grass-green, intermixed with red and violet; while the vent and whole under part of the tail are deep yellow. It seems to have been this latter variety which Buffon describes, who does not appear to have known the genuine colour of the male bird in its complete plumage.

BLACK-CAPPED LORY.

Psittacus Lory. *P. coccineo violaceoque varius, vertice nigro, alis viridibus, cauda viridi rubro violaceoque varia.*

Scarlet and violet Lory, with black crown, green wings, and tail varied with red, green, and violet.

Psittacus Lory. *P. brachyurus purpureus, pileo violaceo, alis viridibus, pectore genubus caudaque cæruleis, orbitis subincarnatis. Lin. Syst. Nat.*

First Black-Capped Lory. *Edw. pl.* 170.

Lori tricolor. *Buff. ois.*

Lory des Philippines. *Pl. Enl.* 168.

Perroquet Lory à scapulaire bleue. *Levaill. pl.* 123. 124.

The size of this beautiful species is nearly equal to that of a common dove-house Pigeon: the crown is violet-black; the cheeks, neck, throat, lower part of the back, and rump scarlet: the upper part of the back, sides, belly, and thighs violet-blue: the tail the same towards the end, the two middle feathers more or less tinged with green: wings green, sometimes slightly edged with scarlet towards the shoulders: bill red; legs and feet black. In the female, according to Monsr. Levaillant, the blue colour is less diffused than in the male, the sides of the body being scarlet. The young male resembles the female, and in both sexes, in the living bird, the bill is red, but fades after death into flesh-colour, or even into white. This species abounds in the Moluccas, and is much esteemed for its gentleness and docility. It is often imported in considerable

COLLARED LORY.

1811, Aug.t 1. London, Published by G. Kearsley, Fleet Street, & the other Proprietors.

numbers to the Cape of Good Hope. During his residence at the Cape Monsr. Levaillant dissected several of these birds, and carefully ascertained their sexual difference as above described. He adds, that it is to this species in particular that the name *Lory*, applied to this tribe of Parrots, belongs; the bird pronouncing that word with extreme distinctness.

COLLARED LORY.

Psittacus Domicella. *P. coccineus, alis viridibus, pileo atroviolaceo, fascia pectorali flava.*

Scarlet Lory, with green wings, violet-black crown, and yellow pectoral bar.

Psittacus Domicella. *P. brachyurus ruber, pileo violaceo, alis viridibus, humeris genibusque cæruleis, orbitis fuscis. Lin. Syst. Nat.*

Second Black-Cap Lory. *Edw. pl.* 171.

Lory à Collier. *Buff. ois.*

Lory des Indes Orientales. *Pl. Enl.* 119. 84.

Perroquet Lori à collier jaune. *Levaill. pl.* 95.

Purple-Capped Lory. *Lath. syn.*

Size of a Dove-House Pigeon: colour scarlet; the wings grass-green, with the ridge of the shoulders blue, and the quill-feathers somewhat dusky towards the tips: across the breast a moderately broad yellow bar, often more or less waved or intermixed with red: thighs violet-blue: crown of the head violet-black; bill yellow or reddish: legs and feet dusky: under coverts of the wings violet-blue; the insides of the quill-feathers yellow with dusky tips: the under surface of the tail less vivid or more inclining to purple than the upper. In some specimens of this beautiful bird the outside edges of the wing-feathers are tinged with yellow; in others the yellow bar on the breast is wanting, and the thighs are green instead of blue.

Like the former, this species is of an active and lively disposition, and remarkable for its docility and distinctness of utterance. It is a native of the Molucca islands.

CERAM LORY.

Psittacus garrulus. *P. coccineus, humeris maculaque dorsali luteis, alis viridibus, cauda versus apicem cærulea.*

Scarlet Lory, with yellow shoulders and back-spot, green wings, and blue-tipped tail.

Psittacus Garrulus. *P. brachyurus ruber, orbitis cinereis, genibus alisque viridibus, rectricibus mediatate postica cærulcis. Lin. Syst. Nat.*

Le Lory de Ceram. } *Buff. ois.*
Le Lory Noira. }

Lori des Moluques. *Pl. Enl.* 216.

Scarlet Lory. *Edw. pl.* 172.

Ceram Lory. *Lath. syn.*

Le Perroquet Lori-Nouara. *Levaill. pl.* 96.

Size of the Collared Lory: colour scarlet, with deep grass-green wings and thighs: shoulder-tips yellow, and at the beginning of the back one or more yellow spots or patches: tips of the wings inclining to dusky or violet-brown: tail generally of the same scarlet colour with the rest of the

RAJA LORY.

1811, Aug.t 1. London, Published by G. Kearsley, Fleet Street, & the other Proprietors.

plumage for about half its length, the remainder blue, but the two middle tail-feathers greenish, and in some specimens the whole colour of this part may rather be termed green than blue. Native of the Molucca isles.

The celebrated Monsr. Levaillant seems to entertain a doubt whether this bird be really distinct, as a species, from the Collared Lory, its general proportions being the same, and the flights of both kinds being observed to associate in their native regions.

RAJA LORY.

Psittacus Raja. *P. coccineus, alis, macula verticali, fascia pectorali, femoribusque luteis.*

Scarlet Lory, with gold-yellow wings, crown-spot, pectoral bar, and thighs.

Lory Rhadia. *Levaill. pl.* 94.

Habit and size of Psittacus Domicella or Collared Lory, of which it is not improbable that it may be a variety. It is most elegantly figured in Monsr. Levaillant's work, and must be allowed to surpass in beauty most of the Lory tribe. Its colour is a vivid scarlet, with the wings entirely gold-yellow: on the top of the head is a broad spot of the same colour, and across the breast a broad bar: the thighs are also yellow: the bill pale or yellowish white, and the legs blackish. This bird is called *Rhadia* in the Moluccas, and is very rare in European collections, and even in its native regions. Monsr. Levaillant, having seen three specimens which exactly agreed in all points, is inclined to consider it as a species distinct from the Collared Lory.

UNICOLOR LORY.

Psittacus unicolor. *P. coccineus, rostro rubro.*
Lory entirely scarlet, with red bill.
Le Perroquet Lori unicolor. *Levaill. pl.* 125.

Size and habit of P. garrulus or Ceram Lory: colour entirely scarlet, deeper or more inclining to crimson on the back and wing-coverts; the very tips alone slightly inclining to purple: bill red: legs and feet dusky, or brownish purple. This bird is finely figured in the work of Monsr. Levaillant, who informs us that it is a native of the Moluccas, and that he has seen two specimens, both in the collection of Mr. Temminck of Amsterdam.

BLUE-TAILED LORY.

Psittacus cyanurus. *P. coccineus, scapularibus caudaque cæruleis, tectricibus cæruleo-variis, remigibus nigricantibus.*
Scarlet Lory, with blue scapulars and tail, wing-coverts varied with blue, and blackish quill-feathers.
Le Perroquet Lori à queue bleue. *Levaill. pl.* 97.

Size and habit of the Collared Lory: colour deep scarlet, or bright crimson; with the scapular feathers and some of the larger wing-coverts fine blue; the rest slightly edged with blue: larger quill-feathers black, slightly edged with blue: tail short in proportion, and blue both above and beneath: vent-feathers blue: bill red; legs black or dusky. Native of Borneo, where it is not uncommon: described from a specimen in the collection of Mr. Raye de Breukelervaert at Amsterdam.

BLUE-FRINGED LORY.

Psittacus cæruleatus. *P. coccineus, cauda sub-purpurea, scapularibus apice cæruleis, tectricibus remigibusque secundariis apice nigris.*

Scarlet Lory, with purplish tail, scapulars tipped with blue, coverts and secondary quills with black. *Nat. Misc. 22. pl. 937.*

Lori à franges bleues. *Levaill. pl. 93.*

Described as a new species by Levaillant: size nearly that of the Black-Capped Lory: colour deep scarlet, with the larger quill-feathers edged, and the smaller pretty deeply tipped with black; as are also the feathers of the alulet or shoulder-edge: middle-coverts fine blue, forming an oblique bar on each wing: insides of the thighs slightly marked with blue: tail very slightly lengthened, and rounded at the tip, but with the feathers considerably shortening on each side from the middle-feathers: the colour of the tail is purplish brown: the bill yellow, and the legs and feet dusky. Native of the Molucca islands, and is not uncommon in the cabinets of the Dutch collectors. The individual described by Levaillant was in that of the Stadholder. In the disposition of its colours it is allied to the Blue-tailed Lory, but the greater length of the tail, together with its different form and colour sufficiently prove it a distinct species.

BLUE-HEADED PARROT.

Psittacus menstruus. *P. viridis, capite collo pectoreque violaceo-cæruleis, crisso caudaque subtus versus basin rubris.*

Green Parrot, with the head, neck, and breast violet-blue; the vent and under parts of the tail towards the base red.

Psittacus menstruus. *P. brachyurus viridis, capite cærulescente, crisso rubro. Lin. Syst. Nat.*

Blue-Headed Parrot. *Edw. pl.* 314.

Perroquet a Camail bleue. *Levaill. pl.* 114.

Papegai à tete et gorge bleue. *Buff. ois.*

Perroquet a tete bleue de Cayenne. *Pl. Enl.* 384.

Blue-Headed Parrot. *Lath. syn.*

FEM.?

Psittacus purpureus. *P. atro-fuscus subtus purpureus, vertice genisque nigris, torque ferrugineo, maculis nigricantibus. Lath. ind. orn.*

Papegai violet. *Buff. ois.*

Perroquet variè de Cayemne. *Pl. Enl.* 408.

Femelle du Perroquet à Camail bleue. *Levaill. pl.* 115.

Little Dusky Parrot. *Edw. pl.* 315. *Lath.*

Rather smaller than a Dove-House Pigeon: head and neck violet-blue, shaded on the breast with purple: remainder of the plumage, in the male, green; but the larger quill-feathers deep blue on the outer edges, and blackish within: vent, and tail-feathers to some distance from the base red: bill ash-coloured, with a reddish or orange-coloured spot on each side the upper mandible: legs blackish. Native of Cayenne, where it is common. The *Little Dusky Parrot* of Edwards, *(Psittacus purpureus Lath.)* is supposed by Levaillant to be the female, which appears to vary much in colour according to its more or less advanced age. That represented by Levaillant being brown, with a strong cast of violet-blue on the wings and under parts; the face blackish, and the head varied with ash-colour: the vent and base of the tail red, as in Edwards's bird.

YELLOW-BREASTED PARROT.

Psittacus Guineensis. ***P. viridis, vertice collo apiceque caudæ rubris, pectore superciliisque flavis, genis gula abdomineque albis.***

Green Parrot, with red crown, neck, and tail-tip, yellow brows and breast, and white cheeks, throat, and belly.

Psittacus Guineensis. ***Miller Ill. pl. 29.***

Yellow-Breasted Lory. ***Lath. syn.***

Size of the preceding species: belly cinereous white: quill-feathers blue with yellowish edges: bill black; legs brown. Said to be a native of Guinea.

WHITE-BREASTED PARROT.

Psittacus melanocephalus. *P. viridis, subtus luteus pectore albido, remigibus apice cæruleis.*

Green Parrot, beneath luteous with whitish breast; the quill-feathers tipped with blue.

Psittacus melanocephalus. *P. brachyurus viridis, subtus luteus, pileo nigro, pectore albo, orbitis incarnatis. Lin. Syst. Nat.*

White-Breasted Parrot. *Edw. pl.* 169. *Lath. syn.*

Le Naïpouri. *Buff. ois. Pl. Enl.* 527. *Levaill. pl.* 119, 120.

Size of a Turtle: breast white, generally waved with dull yellow: bill and legs dusky. Native of Mexico, Guiana, &c.

HOODED PARROT.

Psittacus Caïca. *P. viridis, capite nigro, collo utrinque luteo, macula alari apiceque caudæ cæruleis.*

Green Parrot, with black head, neck luteous on each side, and blue wing-spot and tail-tip.

Psittacus Caica. *P. viridis, capite nigro, orbitis albis, collo flavo, humerorum macula et rectricum apicibus cæruleis. Lath. ind. orn.*

Hooded Parrot. *Lath. syn.*

Le Caïca. *Buff. ois. Pl. Enl.* 744.

Le Perroquet Caïca. *Levaill. pl.* 133.

Var.? Le Caica Barraband. *Levaill. pl.* 134.

Length about eight inches: front of the neck greyish brown: bill and legs reddish. Native of Guiana, where it is known by the name of *Caïca.* In the variety described by Levaillant under the name of *Caïca Barraband*, in honour of the ingenious artist of that name, the cheeks alone are luteous, the breast olive-brown, the inner edges of the shoulders and the bottoms of the thighs orange-colour, and the under wing-coverts red. This, according to Levaillant, forms a distinct race, and is a native of Brasil.

RED-NAPED PARROT.

Psittacus nuchalis. *P. viridis, fronte pectore lunulaque nuchali coccineis.*
Green Parrot, with red front, breast, and nuchal crescent. *Nat. Misc.* 21. *pl.* 913.
Psittacus cervicalis. *P. viridis, fronte lunula cervicis collo subtus pectoreque coccineis. Lath. ind. orn.* Red-Naped Parrakeet. *Lath. syn. supp.*
Le Perroquet Langlois. *Levaill. pl.* 136.

Length about eight inches: bill red: legs brown: supposed by Levaillant to be a native of South-America. Perhaps rather of New-Holland.

GOLDEN PARROT.

Psittacus aureus. *P. totus luteus, rostro pedibusque pallidis.*
Parrot entirely gold-yellow, with pale bill and legs.
Perroquet d'Or. *Levaill. pl.* 138.

Size nearly that of the preceding: length about eight inches: edges of the shoulders slightly tinged with rose-colour. Supposed to be a native of Brasil. Monsr. Levaillant considers this as a truly distinct species, and not a yellow variety of any other kind; which however seems very doubtful.

SENEGAL PARROT.

Psittacus Senegalus. *P. viridis, capite colloque canis, pectore virescente, abdomine flavo.*

Psittacus Senegalus. *P. viridis subtus luteus, capite cinereo, orbitis nigris nudis. Lath. ind. orn.*

Perroquet à tete grise. *Buff. ois. Levaill. pl.* 116, 117.

Petite Perruche du Senegal. *Pl. Enl.* 288.

LENGTH about seven inches: habit stout; tail very short: bill and legs dusky. In the female the grey colour of the head has a greenish cast. Native of Senegal. A variety of this species is mentioned by Levaillant, in which the whole bird is yellow, except the tips of the wings, which are slightly tinged with green.

RED-MASKED PARROT.

Psittacus personatus. *P. viridis, facie gulaque coccineis, vertice subviolaceo, tectricibus inferioribus cæruleis.*

Green Parrot, with scarlet face and throat, subviolaceous crown, and blue under wing-coverts.

Le Perroquet Geoffroy. *Levaill. pl.* 112, 113.

SIZE and habit of the Senegal Parrot: colour grass-green: the top of the head, in the male, pale violet-blue; the forehead, cheeks, and throat scarlet; and the under wing-coverts pale-blue: the bill red, and the legs dusky. The female wants both the blue on the crown and the red on the face. Native of New Holland.

GOLD-WINGED PARRAKEET.

Psittacus chrysopterus. *P. brachyurus viridis, alis macula cærulea fulvaque, orbitis nudis albis. Lin. Syst. Nat.*

Green Parrot, with blue and orange wing-patch, and naked white orbits.

Golden-Winged Parrakeet. *Edw. pl.* 293. *f.* 2. *Lath. syn.*

First described by Edwards: size scarcely larger than that of a Lark: habit rather stout: colour bright-green, somewhat yellower beneath: on each wing a lengthened orange-coloured patch, formed by some of the middle quill-feathers and larger coverts: the larger quill-feathers are blue on the middle of the outside edges, as are likewise the scapulars, thus bounding the orange spot with blue: bill pale, and rather of a more lengthened figure than in most other parrots: orbits bare and whitish: legs dusky. Native of the East Indies.

TOUI PARROT.

Psittacus Toui. *P. viridis, macula gulari alarique fulvis.*
Green Parrot, with a fulvous spot on the throat and wings.
Psittacus Toui. Yellow-Throated Parrakeet. *Lath.*
Toui à gorge jaune. *Buff. ois.*
Petite Perruche à gorge jaune? *Pl. Enl.* 190. *f.* 1.
La Perruche à tache souci? *Levaill. pl.* 58, 59.

Size of the Gold-Winged Parrot: immediately under the throat a roundish subfulvous spot: on the wing-coverts a chesnut-fulvous bar: bill grey: legs dusky.

Monsr. Levaillant ranks his *Perruche à tache souci*, which I suspect to be the same species, among the former division of the genus, the tail-feathers (in his specimen) being pointed, though short; but this is the case in several of the short-tailed Parrots. In Monsr. Levaillant's figure the fulvous wing-patch is also succeeded by a blue one, the outsides of some of the quill-feathers, as well as the two middle tail-feathers being of that colour. Native region uncertain.

TUI PARRAKEET.

Psittacus Tui. *P. viridis, fronte aurantia, orbitis flavis, rostro nigro. Lath. ind. orn.*
Green Parrot, with orange forehead, yellow orbits, and black bill.
Gold-Headed Parrakeet. *Lath. syn.*
La Toui à tete d'or. *Buff. ois.*
Petite Perruche de l'isle de St. Thomas. *Pl. Enl.* 456. *f.* 1.
La Perruche Tui. *Levaill. pl.* 70.

Said by Brisson to be of the size of a Starling; but the specimens figured in the Planches Enluminées and the work of Levaillant far less, being not larger than the Otaheitan Blue Parrakeet: colour entirely green, except the forehead, pretty deep on the crown, which is pale orange: head and neck slightly tinged with a cast of pale blue: bill and legs yellowish-brown. Native of Cayenne, where, according to Levaillant, it is common.

PURPLE-TAILED PARROT.

Psittacus porphyrurus. *P. viridis, uropygio cyaneo, rectricibus (exceptis intermediis duabus) purpureis. Vivar. Nat.* 1. *pl.* 16.
Green Parrot, with the rump blue, and the tail (except the two middle feathers), purple. *Nat. Misc. vol.* 1. *pl.* 16.
Psittacus purpuratus. Purple-tailed Parrakeet. *Lath.*

Size of a Lark: length between seven and eight inches: crown and back dusky green; remainder bright-green: edges of the shoulders and tips of the wings dusky: tail-coverts lengthened and fibrous; bill pale; legs dusky. Native of Cayenne; a rare and highly elegant species.

BLACK-WINGED PARRAKEET.

Psittacus melanopterus. *P. thalassinus, dorso alisque nigris, remigibus secundariis flavis apice cæruleis, cauda purpurea fascia nigra.*

Sea-green Parrot, with black wings; the secondary quill-feathers yellow with blue tips; the tail lilac-coloured, crossed by a black bar.

Psittacus melanopterus. Black-winged Parrakeet. *Lath. Nat. Misc.* 4. *pl.* 132. *Brown Ill. pl.* 8.

Perruche à ailes variées. *Buff. ois.*

Petite Perruche de Batavia. *Pl. Enl.* 791. *f.* 1.

La Perruche Javane. *Levaill. pl.* 69.

A BEAUTIFUL species: size smaller than that of the immediately preceding: forehead and back of the neck slightly tinged with yellow: tail-feathers pointed, and of a beautiful blueish purple or lilac-colour, crossed near the tip by a black bar: bill pale; legs brown. Native of Java.

BLACK-WINGED PARRAKEET.

M. Griffith sculp.

GUINEA PARROT.

GUINEA PARROT.

Psittacus pullarius. *P. viridis facie rubra, uropygio cæruleo, cauda fulva fascia nigra.*

Green Parrot, with red face, blue rump, and orange-red tail crossed by a black bar.

Psittacus pullarius. *P. brachyurus viridis, fronte rubra, cauda fulva fascia nigra, orbitis cinereis. Lin. Syst. Nat.*

Little Red-Headed Parrakeet, or Guinea Sparrow. *Edw. pl.* 237. *Lath. syn.*

Guinea Parrakeet. *Nat. Misc. vol.* 1. *pl.* 35.

La Perruche à tete rouge de Guinèe. *Buff. ois.*

Petite Perruche male de Guinèe. *Pl. Enl.* 60.

Var? Psitt: minor. Amer: versicolor. *Seb.* 2. *t.* 40. *f.* 1.

A HIGHLY beautiful species: length about five inches and a half: habit stout: bill orange; legs grey. Native of Guinea, where it is common. It is remarkable for the gentleness of its disposition, but, in general, cannot be taught to articulate. There can be little doubt that the figure in *Seba's Thesaurus*, vol. 2. pl. 40. f. 1. is intended for the present species. What seems to have rendered this doubtful to Buffon is the declaration of Seba relative to the distinct articulation and imitative powers of the individual there described.

GREY-HEADED PARROT.

Psittacus canus. *P. viridis, capite colloque griseis, cauda fascià nigra.*

Green Parrot, with grey head and neck, and black bar across the tail.

Grey-Headed Parrakeet. *Lath. syn.*

La Perruche à tete grise. *Buff. ois.*

Petite Perruche de Madagascar. *Pl. Enl.* 791. *f.* 2.

Size of a Common Sparrow: bill and legs pale: tail-feathers pointed, as in the Black-Winged Parrot, though the tail, when closed, appears rounded. Native of Madagascar.

SAPPHIRE-CROWNED PARROT.

Psittacus Galgulus. *P. viridis, uropygio pectoreque coccineis, vertice cæruleo. Lin. Syst. Nat.*

Green Parrot, with red rump and breast, and blue crown.

Sapphire-Crowned Parrakeet. *Edw. pl.* 293. *f.* 2.

La Perruche à tete bleue. *Buff. ois.*

Petite Perruche de Perou. *Pl. Enl.* 190. *f.* 2.

VAR.

Philippine Parrakeet. *Lath. syn.*

Perruche des Philippines. *Pl. Enl.* 520.

Le Coulacissi. *Buff. ois.*

A most elegant species; size about a third less than the Guinea Parrot: bill and legs grey: in some specimens, perhaps young birds or females, the blue spot on the crown is wanting. Native of the Philippine islands.

COLLARED PARROT.

Psittacus torquatus. *P. viridis, fascia nuchæ lutescente nigro undulata. Lath. ind. orn.*

Green Parrot, with a broad yellow bar undulated with black at the back part of the neck.

Le Perruche à Collier. *Buff. ois.*

Petite Perruche à Collier de l'isle de Luçon. *Sonn. Voy. p. 77. pl. 39.*

Collared Parrakeet. *Lath. syn.*

Size of the Guinea Parrot: bill and legs dark grey: tail short and pointed. In the female the collar at the back of the head is pale blue instead of yellow, but spotted or waved with black, as in the male. Native of the Philippine islands.

CRESTED RED PARROT.

Psittacus erythrochlorus. *P. viridis cristatus, occipite alis caudaque rubris. Lath. ind. orn.*

Green Parrot, with red hind-head, crest, wings, and tail.

Crested Red and Green Parrakeet. *Lath. syn.*

Described by Aldrovandus: said to be of the size of a Blackbird: the crest consists of six feathers, three of which are longer than the rest.

SOSOVÈ PARROT.

Psittacus Sosovè. *P. viridis, macula alari uropygioque flavis, orbitis nudis albidis, rostro pedibusque carneis.*
Green Parrot, with yellow wing-spot and rump, naked whitish orbits, and flesh-coloured bill and legs.
Cayenne Parrakeet. *Lath. syn.*
Sosovè. *Buff. ois.*
Petite Perruche de Cayenne. *Pl. Enl.* 456.*f.* 2.

ABOUT the size of the Guinea Parrot. Native of Cayenne, where it is called Sosovè, and is said to be of a very docile disposition.

TIRICA PARROT.

Psittacus Tirica. *P. viridis, rostro incarnato, pedibus cærulescentibus. Lath. ind. orn.*
Green Parrot, with flesh-coloured bill and blueish legs.
Green Parrakeet. *Lath. syn.*
Le petite Jaseuse. *Pl. Enl.* 837.
Tirica. *Buff. ois.*

ABOUT the size of the Guinea Parrot: said to be a native of Brasil.

LUZONIAN PARROT.

Psittacus minor. *P. viridis, pileo tectricibusque caudæ coccineis, pectore cæruleo. Lath. ind. orn.*

Green Parrot, with scarlet crown and rump, and blue breast.

La Perruche à ailes noires. *Buff. ois.*

Petites Perruches de l'isle de Luçon. *Sonner. Voy. p. 77. pl. 4.*

Luzonian Parrakeet. *Lath. syn.*

Smaller than the Guinea Parrot: larger quill-feathers black: bill and legs yellow. In the female there is only a red spot immediately above the upper mandible. Native of the island of Luzonia.

ASIATIC PARROT.

Psittacus Asiaticus. *P. viridis, rostro pileo dorsoque postico fulvis, remigibus rectricibusque subtus cæruleo-viridibus, orbitis pedibusque incarnatis. Lath. ind. orn.*

Green Parrot, with fulvous-red bill, crown, and rump; wings and tail blue-green beneath, and flesh-coloured orbits and legs.

Red and green Indian Parrot. *Edw. pl. 6.*

Size somewhat smaller than that of the Guinea Parrot. Native of India, and seems much allied to the immediately preceding species.

CHESNUT-CROWNED PARROT.

Psittacus Anaca. *P. viridis, subtus rufo-fuscus, pileo castaneo, gula cinerea, macula dorsali cauduque pallidè fuscis, margine alarum rubra. Lath. ind. orn.*

Green Parrot, rufous-brown beneath, with chesnut crown, cinereous throat, pale-brown back-spot and tail, and red wing-edge.

Chesnut-Crowned Parrakeet. *Lath. syn.*

Anaca. *Will. orn. Raii syn.*

Size of a Lark: greater quill-feathers tipped with sea-green: bill, orbits, and legs dusky. Native of Brasil and Guiana.

BLUE-RUMPED PARROT.

Psittacus Malaccensis. *P. viridis, fronte uropygioque cæruleis, tectricibus alarum inferioribus rubris. Lath. ind. orn.*

Green Parrot, with blue forehead and rump, and red under wing-coverts.

Blue-Rumped Parrakeet. *Lath. syn. suppl.*

Petite Perruche de Malacca. *Sonner. voy. Ind. 2. p. 212.*

Size of the Guinea Parrot? bill violet-grey; legs brown: irides red: tail yellowish beneath. Native of Malacca.

TUIPARA PARROT.

Psittacus Tuipara. *P. viridis, lunula frontis rubra, macula alarum lutea, rostro incarnato. Lath. ind. orn.*

Green Parrot, with red frontal crescent, luteous wing-spot, and flesh-coloured bill.

Red-Fronted Parrot. *Lath. syn.*

Tuipara. *Will. orn. Raii syn.*

Size of a Lark: tail very short: legs grey. Native of Brasil.

FRINGILLACEOUS PARROT.

Psittacus fringillaceus. *P. viridis, capite cyaneo, genis gula jugulo maculaque abdominis ex lacteo sanguineis, abdomine violaceo. Lath. ind. orn.*

Green Parrot, with blue head, pale red cheeks, throat and belly-spot, and violaceous abdomen.

Sparrow Parrot. *Lath. syn. suppl.* 2.

Size of a Sparrow: length six inches: bill and legs yellow: insides and tips of the tail-feathers yellow. Native of South America.

PASSERINE PARROT.

Psittacus passerinus. *P. viridis, macula alarum uropygio alisque subtus cæruleis.*

Green Parrot, with blue wing-patch and rump, and wings blue beneath.

Least blue and green Parrakeet. *Edw. pl.* 235.

Psittacus passerinus. *P. brachyurus luteo-virens, macula alarum alisque subtus cæruleis. Lyn. Syst. Nat. Lath. ind. orn.*

Tuietè. *Will. orn. Raii syn.*

This is, perhaps, the smallest of the short-tailed Parrots, measuring but little more than four inches in length: colour green, somewhat yellowish beneath: on the wings a blue patch: rump and upper tail-coverts pale blue: under surface of the wings, according to Edwards, greenish ash-colour, mixed with blue on the smaller coverts; but, according to Linnæus, blue. Native of Brasil and Guiana.

VAR.?

CAPE PARROT.

Psittacus Capensis. *P. viridis, alis supra cæruleo variis, subtus cæruleis, rostro pedibusque rubescentibus. Lath. ind. orn.*

Green Parrot, with the wings varied with blue, and blue beneath, and the bill and legs reddish.

Perruche à ailes bleues. *Buff. ois.*

Petite Perruche de Cap de Bonne Experance. *Pl. Enl.* 445. *f.* 1.

Blue-Winged Parrakeet. *Lath. syn.*

Size similar to that of the preceding, of which it is, perhaps, rather a variety than a truly distinct species: plumage green, but some of the wing-feathers blue: under wing-coverts blue: bill and legs reddish. Native, according to Buffon, of the Cape of Good Hope. It is said sometimes to vary in having a blue spot on the crown, and yellow legs.

END OF THE EIGHTH VOLUME.

T. DAVISON, Lombard-street,
Whitefriars, London.

Made in United States
Troutdale, OR
11/03/2026

56844529R00166